THE DESIGN AND APPLICATION OF

PROGRAMMABLE SEQUENCE CONTROLLERS FOR AUTOMATION SYSTEMS

D. Pessen

Associate Professor, Department of Mechanical Engineering, Technion, Haifa, Israel

W. Hübl

Wissenschaftlicher Assistent, Institut für Automatisierungstechnik, Ruhr-Universität Bochum, Bochum, German Federal Republic

Longman
London and New York

Longman Group Limited London

Associated companies, branches and representatives throughout the world

Published in the United States of America by Longman Inc., New York

First published 1979

British Library Cataloguing in Publication Data

Pessen, D
The design and application of programmable sequence controllers for automation systems.
1. Sequence controllers
I. Title II. Hübl, W.
629.8'91 TJ223.S/ 78-40456

ISBN 0-582-50401-5

Set in Compugraphic Times 11 on 12pt
and printed in Great Britain by
Richard Clay (The Chaucer Press) Ltd,
Bungay, Suffolk

CONTENTS

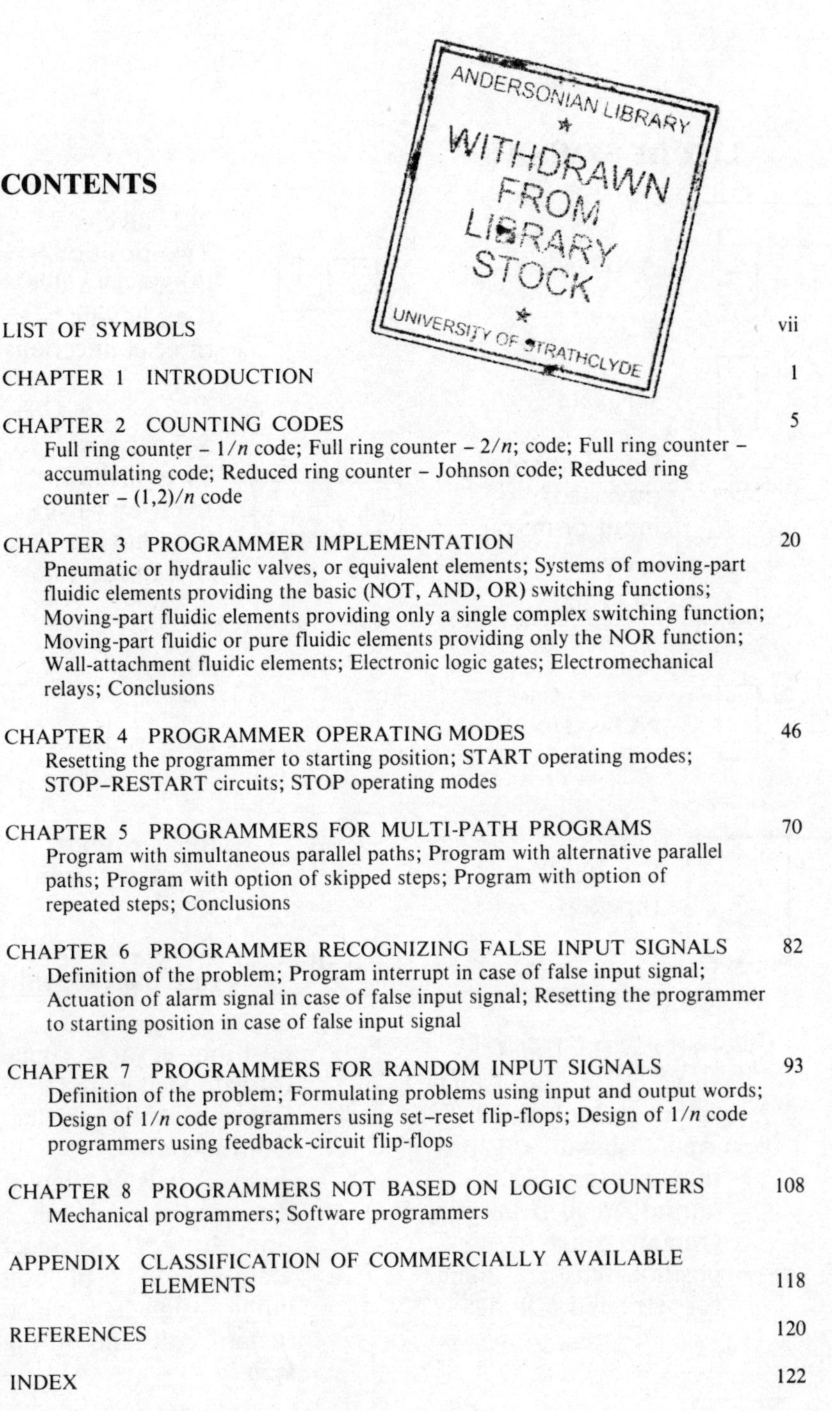

LIST OF SYMBOLS

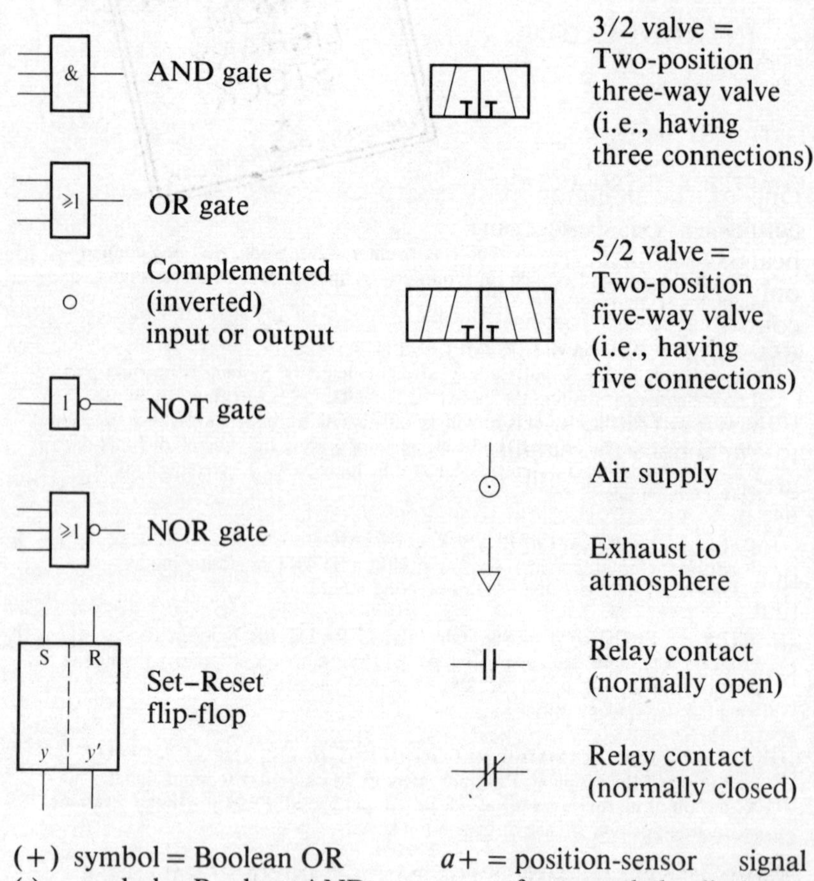

(+) symbol = Boolean OR

(·) symbol = Boolean AND

(′) symbol = Boolean NOT

x_i = input signal to prorammer stage i

z_i = output signal from programmer stage i

$a-$ = position-sensor signal for retracted cylinder A

$a+$ = position-sensor signal for extended cylinder A

k_i = Boolean AND product (conjunction) of all input signals existing at the completion of step i

k_i^* = Boolean AND product (conjunction) of all input signals which remain constant during step i

CHAPTER 1

INTRODUCTION

One of the common problems of industrial automation is to actuate various devices, such as cylinders, pumps, motors, timers, heaters, etc. in a certain sequence. Each actuation must take place only after the previous step has been completed. The necessary control systems for handling problems of this type are called *sequential systems*.

Originally, sequential systems were designed mainly by intuition, using either electrical relays or pneumatic or hydraulic valves to implement the circuit. Intuitive methods, however, depend a great deal on the skill and cleverness of the designer, and their use becomes impractical for large systems.

One systematic method for designing sequential systems is that originally developed by Huffman [1]*, and described in virtually every textbook on switching theory. This method utilizes so-called primitive flow tables, merged flow tables, and finally Karnaugh maps to design, what is called, an asynchronous system. The method results in a set of binary-variable or Boolean equations, which can then be physically implemented by means of any desired type of logic elements. While this method requires little intuitive ability, it becomes unwieldy and impractical for even moderately complicated programs, because the number of binary variables quickly becomes too large to handle. In addition, the method demands that the designer be fairly skilled in the techniques of switching theory. And a final and perhaps biggest drawback: the solution obtained holds only for one specific program. If even a slight change is made in this program, we usually have to start again from scratch and work out a completely new solution.

A different design method, first developed by Cole and Fitch [2], [3], [4], uses so-called synthesis tables, and will often provide

* Roman numerals in [brackets] indicate References at the end of the book, on p. 120. Superior numerals indicate Notes at the end of chapters.

a more economical solution than Huffman's method, especially where pneumatic valves are used as switching elements. (The method is not economical for relays.) Though large-scale problems can be handled with this method, quite a bit of design work is still required, and, as with Huffman's method, changes in the program will require a new design.

A completely different approach to the problem is to use one of the various types of commercially available sequential-system programmers. These can be in the form of punched-card or punched-tape readers (both electrical and pneumatic versions of such readers are available) or stepping-switch drum-type programmers driven by stepping motors. While these systems are extremely flexible – the program is changed simply by changing the tape, card or drum – they are rather expensive, and their use can be economically justified only where very long programs or frequent changes are required. Also, such systems are subject to the wear-and-tear of all mechanical devices.

A purely electronic solution is provided by the programmable logic controller, which is, in effect, a specialized mini-computer. These devices, which have become widely accepted in industry during the last few years, provide the utmost in flexibility and convenience of operation. Though their cost has dropped drastically, they are still too expensive to use for small or medium-length programs.

The purpose of this book is to discuss in detail so-called *logic counter programmers*. These are programmable sequence controllers which the user can easily assemble himself from standard commercially-available logic components (be it pneumatic or hydraulic valves, moving-part fluidic elements, pure fluidic elements, electronic gates, or electromechanical relays). The programmer can thus be completely compatible with the type of logic elements that are being used in any particular plant. Such programmers combine low cost with the flexibility and ease of programming of more expensive ready-made programmers. The main advantages of logic counter programmers are as follows:

1. The programmer consists of identical standard modules, and is therefore very easy to put together and connect. Each module is built from the same standard off-the-shelf logic elements.
2. The programmer is suitable for sequential programs of any desired length. By adding sufficient modules, the programmer can be made to accommodate even the most complicated program.

3. Connecting the programmer to the system to be controlled (i.e., interfacing) requires no special skill. The programmer appears to the user as a 'black box' with a number of input connections $x_1 \ldots x_n$ and an equal number of output connections $z_1 \ldots z_n$, where n is the number of program steps. For any given program step i, we connect z_i to the outside element to be actuated during that step, and x_i to the sensor announcing that the step has been carried out.
4. Changing the program does not require a complete redesign of the control system. Program changes are easily made simply by changing the appropriate input (x_i) and output (z_i) connections.
5. Because of its extreme simplicity, it becomes very easy to explain the method to maintenance and operating personnel, which facilitates trouble-shooting and maintenance work. Trouble-shooting becomes especially easy. Since each program step is associated with one and only one programmer module, we know exactly where to search if a certain step is not carried out properly.

While logic counter programmers often require somewhat more logic elements than a sequential switching system especially designed for the particular problem, this does not necessarily mean that the overall cost will be higher. The engineering time saved during the design stage, the ease of making program changes and the convenience in trouble-shooting can more than make up for the cost of the additional logic elements.

Because of their many advantages, logic counter programmers have come into wide use in continental Europe[1], but somehow they are little known in English-speaking countries. Their many advantages make it clearly worthwhile to use them wherever applicable, and it is hoped that this book will help to make this versatile method more widely known and used. In Chapter 2, the various counting codes that have been suggested for such programmers in the literature are critically compared. In Chapter 3, different types of logic elements are discussed, and the most suitable counting codes for each particular type of element are pointed out. Chapter 4 discusses various programmer operating modes (programmer reset, Start modes, Stop modes, etc.) and their necessary control units, while Chapter 5 treats programmers for multi-path programs, again in conjunction with the various possible counting codes. Chapter 6 discusses the design of programmers able to recognize false input signals, a valuable safety feature in critical applications. Finally, Chapter 7 describes a new design method for programmers that can accommodate programs with random (optional) input signals. In Chapter 8, other types of programmers are briefly discussed.

It is assumed that the reader is familiar with the rudiments of switching (Boolean) algebra. The algebraic expressions needed here are generally very simple, except for Chapter 7, where, due to the nature of the topic discussed, lengthy equations are required.

A thorough search of the available technical literature in this field was made, and all relevant material found has been included here. Nevertheless, more than half of the material presented here is new, and many of the new circuits shown are felt to represent a considerable advance in the state of the art. In order to assure the reliability of these new circuits, they have all been checked out on a laboratory test bench. Thus, the user can apply them to his specific problem with complete confidence.

Notes

1. In Germany, they are known by any one of the following names: Taktkette, Schiebetaktkette, Taktstufensteuerung, Schrittregister, Schrittschaltkette, and even others.

CHAPTER 2

COUNTING CODES

Logic counter programmers keep track of the various program steps by counting them; hence the name counter programmer. The counters used are so-called ring counters, i.e., after counting the final step, the counter automatically returns to Step No. 1. Counters use memory elements (usually set–reset flip-flops) to keep track of the count. The most economical ring counter, from the standpoint of number of flip-flops required, would use the natural binary code. Since each flip-flop has two possible output states, namely $y=0$ or $y=1$, such a counter can count 2^n steps with n flip-flops, as illustrated in Table 2.1 (for $n=3$).

Table 2.1 Ring counter using natural binary code

Step No.	y_1	y_2	y_3
1	0	0	1
2	0	1	0
3	0	1	1
4	1	0	0
5	1	0	1
6	1	1	0
7	1	1	1
8	0	0	0

However, programmers based on counters of this type lose some of the advantages listed previously: they do not consist of standard modules, and therefore are less flexible in use. Furthermore, the *decoding* (i.e., deciding which step we are at) becomes more complicated, which cancels the benefit of needing less flip-flops. We shall therefore investigate other codes which are simpler, even though they may require more flip-flops.

A. Full ring counter – 1/*n* code

The first three codes to be described in the following

discussion represent *full ring counters*. In a full ring counter, each program step is assigned one flip-flop. The simplest possible code for a full ring counter is shown in Table 2.2 (again for a program of eight steps, although the program can, of course, be expanded indefinitely). The code will be called the $1/n$ ('one-out-of-n') code since, at any given instant, one and only one out of the n flip-flops will be in the SET condition. The $1/n$ code is the most common code used for logic counter programmes. Indeed, most of the applications reported in the literature use this code; see, for example [5], [6], [7].

Table 2.2 $1/n$ code

Step No.	y_1	y_2	y_3	y_4	y_5	y_6	y_7	y_8
1	1	0	0	0	0	0	0	0
2	0	1	0	0	0	0	0	0
3	0	0	1	0	0	0	0	0
4	0	0	0	1	0	0	0	0
5	0	0	0	0	1	0	0	0
6	0	0	0	0	0	1	0	0
7	0	0	0	0	0	0	1	0
8	0	0	0	0	0	0	0	1

A programmer based on the $1/n$ code is illustrated in Fig. 2.1. Only four stages are shown in the figure. The remaining stages (as many as may be required for the program at hand) are connected in identical fashion. Each standard module consists of one flip-flop and one AND gate. The AND gate provides the SET signal for the flip-flop i provided the previous flip-flop $(i-1)$ is in the SET condition, AND provided the signal x_{i-1} is actuated,

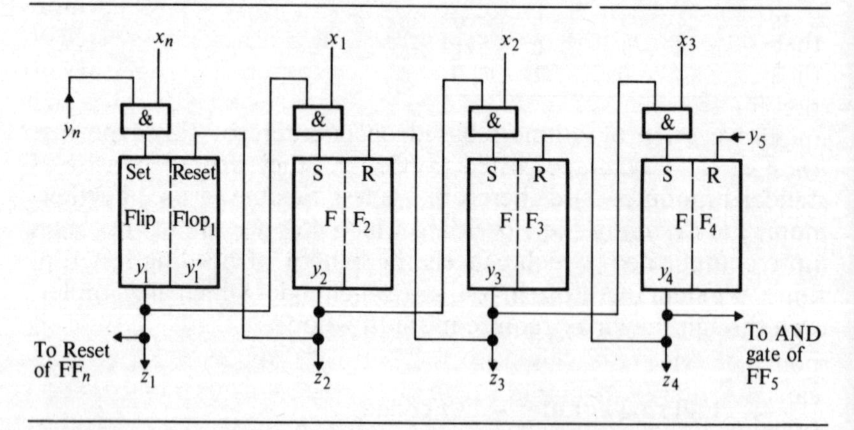

Fig. 2.1 Programmer based on $1/n$ code of Table 2.2 (four stages)

indicating that the previous program step has been completed. The moment a given flip-flop i is placed in the SET condition, its y output does three things:

1. Resets the previous flip-flop $(i-1)$.
2. Switches the signal $z_i = 1$, thus actuating the required external equipment.
3. Sends a 1 signal to the AND gate of the succeeding flip-flop $(i+1)$, thus getting this flip-flop ready for its SET signal the moment x_1 becomes 1.

In a typical system where the external equipment to be controlled consists of a number of pneumatic or hydraulic cylinders, the various x signals will usually be provided by mechanically actuated 3-way limit valves with spring return, or by non-contacting fluidic sensors, or by limit switches in the case of relay or electronic control systems. The various z signals are the pilot signals for the valves actuating the various cylinders. Note that the AND gates are required to avoid an improper count in case an x signal should be operated out of turn, or where a given x signal appears more than once during the program cycle. By using these AND gates, we make sure that a given flip-flop i will only be set if the cycle has reached the previous $(i-1)$ step.

If two (or more) cylinders must be actuated simultaneously during a step, the x signal signifying that the step has been completed is obtained from an AND gate whose inputs are the appropriate limit valves or switches. In this way, the next step is not initiated until all of the cylinders have completed their strokes of the previous step.

In cases where a given pilot signal has to be operated more than once during the cycle, the pilot line is usually connected to the required z lines through an OR gate. This is necessary to prevent false actuation of other AND gates or RESET lines that might occur if more than one z line were directly connected to the same pilot line without first going through an OR gate.

Suppose a given output signal has to be maintained continuously for a number of steps $(=n)$ (e.g., continuous operation of a motor or heater, or continuous pilot signal to an actuating valve with spring return). We could connect the n succeeding z signals $z_i \ldots z_{i+n-1}$ through an OR gate to the required output point. However, for large n, this will probably exceed the fan-in capacity of the available OR gates, which would require additional OR gates connected in a branched manner. To avoid this, it will usually be more convenient to use an additional flip-flop, whose output is connected to the required output point, as done in [5]. The z_i signal will set the flip-flop, while the RESET signal is provided by z_{i+n}. The choice is made purely on economic grounds; i.e., what costs more: the flip-flop, or the required number of OR gates with associated tubing or wiring.

B. Full ring counter – 2/*n* code

An alternative code for a full ring counter is suggested in Table 2.3, and four stages of the resulting programmer, first suggested in [8], are shown in Fig. 2.2. In the 2/*n* code, there is

Table 2.3 2/*n* code

Step No.	y_1	y_2	y_3	y_4	y_5	y_6	y_7	y_8
1	1	1	0	0	0	0	0	0
2	0	1	1	0	0	0	0	0
3	0	0	1	1	0	0	0	0
4	0	0	0	1	1	0	0	0
5	0	0	0	0	1	1	0	0
6	0	0	0	0	0	1	1	0
7	0	0	0	0	0	0	1	1
8	1	0	0	0	0	0	0	1

overlap in the flip-flop actuation; i.e., any given flip-flop remains set during two consecutive steps. This means that two flip-flops will be in the set condition during any step; hence the name 2/*n* code. This code is especially economical where pneumatic valves

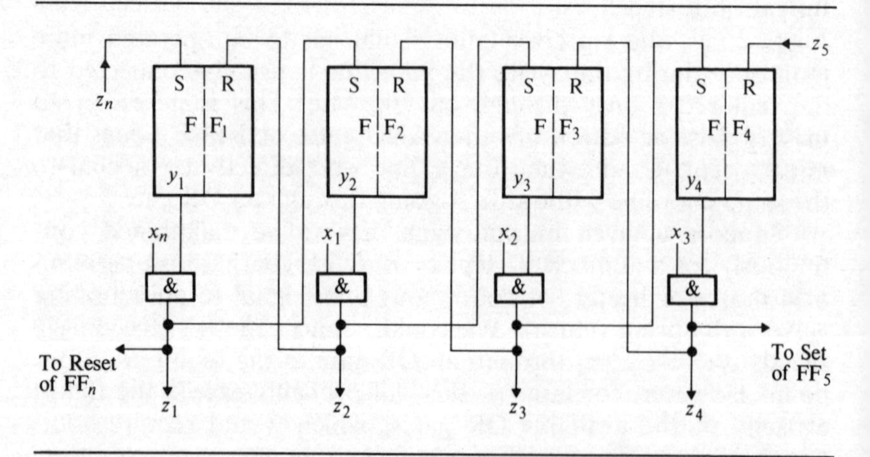

Fig. 2.2 Programmer based on 2/*n* code of Table 2.3 (four stages)

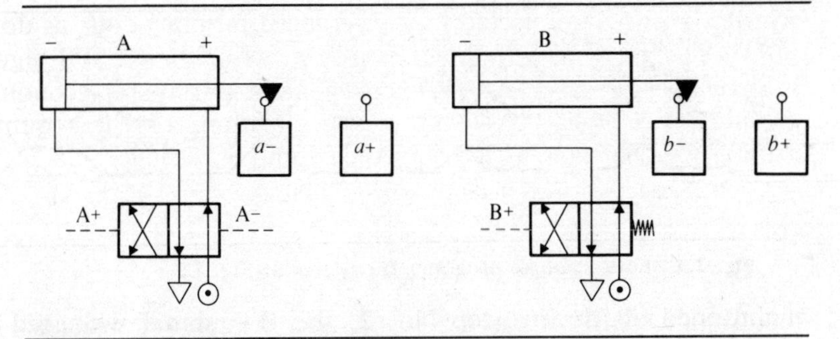

Fig. 2.3 Actuation of two pneumatic cylinders

are used, as will be shown in Chapter 3. The code also permits us to eliminate the OR gate where a pilot line must be actuated for two consecutive steps. For example, if we wished to actuate a given pilot line during steps No. 2 and 3, we could connect this pilot line directly to flip-flop output y_3 (rather than to z_2 and z_3 through an OR gate), since Table 2.3 shows us that y_3 will be 1 during steps No. 2 and 3. However, the OR gate or a flip-flop will still be required if a given pilot line has to remain actuated during more than two consecutive steps.

Programmers according to the $2/n$ code have one limitation: the input signals x_i must be sustained signals, if we wish the corresponding output signals z_{i+1} to be sustained rather than only short pulses. This is clear from Fig. 2.2: since x_i and z_{i+1} are input and output respectively of the same AND gate, the z_{i+1} signal will more or less duplicate the shape of the x_i signal. (This limitation does not exist with the $1/n$ code. As can be seen from Fig. 2.1, a pulse x_i is sufficient to set the flip-flop, producing a sustained z_{i+1} signal.)

The above limitation may cause a problem in cases where cylinder actuating valves with spring return are used, since these require sustained pilot signals. As an example, consider the two cylinders shown in Fig. 2.3. Cylinder A is actuated by a valve with two pneumatic pilot inputs (i.e., memory valve), while the actuating valve for cylinder B has one pilot input acting against a return spring. The time–motion diagrams of Fig. 2.4 illustrate several instructive hypothetical situations, each of which should be considered as part of a larger program.

In the case of Fig. 2.4(a), there is no problem. At the end of step No. 1, the limit valve a+ is actuated. This provides the x_1 signal in the programmer of Fig. 2.2, giving the z_2 output signal which is connected to the pilot line B+. Since the a+ signal is

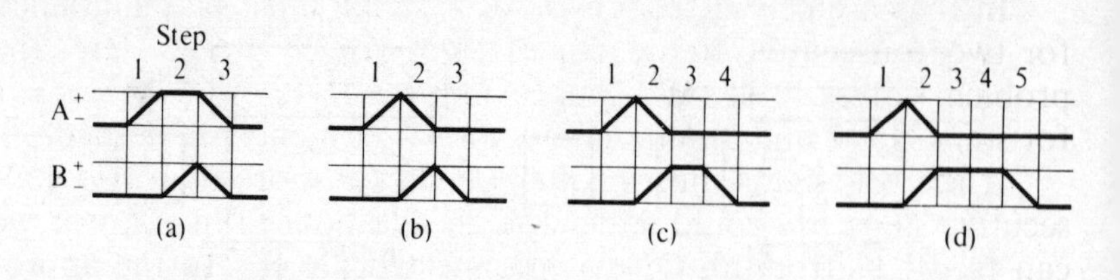

Fig. 2.4 Four example programs for system of Fig. 2.3

maintained all during step No. 2, the B+ signal will also be maintained.

The situation is different for the case of Fig. 2.4(b). Here, cylinder A returns immediately to the (−) position, so that the a+ signal will only be a short pulse. In order to obtain a B+ signal which remains sustained all during step No. 2, we need an additional AND gate connected to the outputs of flip-flops 2 and 3. Rather than connecting the B+ pilot line to the z_2 signal, we connect it to this AND gate, giving B+ $= y_2y_3$. Another solution would be to use an additional flip-flop, rather than an additional AND gate, with $a+$ as SET and $b+$ as RESET signals, and with the flip-flop output providing the B+ signal. However, a flip-flop will, in general, cost more than an AND gate.

Table 2.4 Single- and multi-state decoding for 2/*n*-code programmer, showing necessary gate inputs to provide sustained output signals for *m* consecutive steps

m	Sustained output signal for steps No. i to $i+m-1$	Output function for sustained x signals	Output function for pulsed x signals, with y variables being	
			accessible	not accessible (see Fig. 3.7)
1	Step No. i	z_i	$y_i \cdot y_{i+1}$	Use flip-flop
2	Steps No. i to $i+1$	y_{i+1} or $z_i + z_{i+1}$	y_{i+1}	Use flip-flop
3	Steps No. i to $i+2$	$y_{i+1} + y_{i+2}$ or $z_i + z_{i+1} + z_{i+2}$	$y_{i+1} + y_{i+2}$	Use flip-flop
4	Steps No. i to $i+3$	$y_{i+1} + y_{i+3}$ or $z_i + z_{i+1} + z_{i+2} + z_{i+3}$	$y_{i+1} + y_{i+3}$	Use flip-flop
5	Steps No. i to $i+4$	$y_{i+1} + y_{i+3} + y_{i+4}$	or flip-flop	Use flip-flop
6	Steps No. i to $i+5$	$y_{i+1} + y_{i+3} + y_{i+5}$	or flip-flop	Use flip-flop

In the case of Fig. 2.4(c), the B+ signal must be maintained for two consecutive steps, namely steps No. 2 and 3. Here, the problem solves itself. We simply connect B+ $= y_3$ (since $y_3 = 1$ for steps No. 2 and 3), and no additional elements are needed.

If the B+ signal must be maintained for more than two consecutive steps, we could either use an additional flip-flop, or we can feed B+ from an OR gate connected directly to the appropriate flip-flop outputs. This procedure is called *multi-state decoding*. For example, if the B+ signal must be maintained during steps No. 2, 3 and 4, as shown in Fig. 2.4(d), we would use B+ $= y_3 + y_4$. (This would be more economical than using B+ $= z_2 + z_3 + z_4$, even for cases where the z signals, being sustained, could be used.) Table 2.4 summarizes the results. The table also covers cases where the output signal must be maintained for $m > 3$ consecutive steps. As can be seen from the table,

for $m =$ even, an OR gate with fan-in $= m/2$ is required;
for $m =$ odd, an OR gate with fan-in $= (m + 1)/2$ is required;

unless we choose to use a flip-flop instead.

C. Full ring counter – accumulating code

A third type of full ring counter uses the accumulating code described in Table 2.5. This code is so called because, for each step, an additional 1 is accumulated. At the end of the cycle, all of the flip-flops are reset simultaneously, and the programmer is then ready for the next cycle. Programmers based on this cycle are discussed in [9].

Table 2.5 Accumulating code

Step No.	y_1	y_2	y_3	y_4	y_5	y_6	y_7	y_8
1	1	0	0	0	0	0	0	0
2	1	1	0	0	0	0	0	0
3	1	1	1	0	0	0	0	0
4	1	1	1	1	0	0	0	0
5	1	1	1	1	1	0	0	0
6	1	1	1	1	1	1	0	0
7	1	1	1	1	1	1	1	0
8	1	1	1	1	1	1	1	1

The programmer based on the accumulating code is shown in Fig. 2.5. It is quite similar to that of Fig. 2.1 based on the $1/n$ code, except that the output signals z_i do *not* provide the RESET

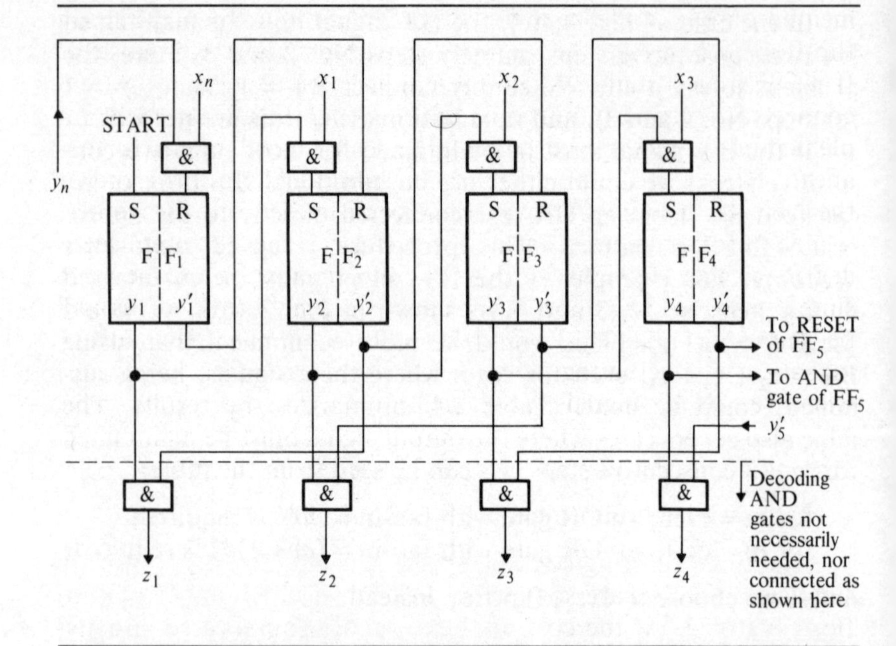

Fig. 2.5 Programmer based on accumulating code of Table 2.5 (four stages)

signals for the previous flip-flops. When the cycle is completed, the last flip-flop is set, and the resulting y_n signal together with x_n resets the first flip-flop, provided the START signal is interrupted. Each complemented flip-flop output y_i' provides the RESET signal for the next flip-flop FF_{i+1}, so that all the flip-flops are successively reset. A START signal is then needed to initiate step No. 1, so that the cycle will start anew. (It would seem possible to have the $y_n x_n$ signal directly actuate the RESET signals of all the flip-flops simultaneously. However, this would necessitate inserting a delay in the RESET line of the last flip-flop so that y_n does not become 0 before all the remaining flip-flops have been reset.)

In the above programmer, each y_i, once actuated, remains 1 until the end of the cycle. This is of advantage if we need sustained output signals z_i, since it eliminates the need for the additional AND or OR gates or flip-flops that were mentioned in connection with the $1/n$ and $2/n$ codes. However, whenever we want to cancel the z_i signal, this fact becomes a drawback, since it forces us to add additional 'decoding' AND gates to provide the z outputs, as shown in Fig. 2.5. In the figure, each such AND gate is shown with its second input connected to the complemented output of the next flip-flop. Thus, the moment a certain flip-flop becomes set, the previous *z* is automatically extinguished. If we wish to sustain a given *z* signal for several steps, we would connect the second input of its decoding AND gate to the complemented output of the correct flip-flop several stages further ahead. Finally, if the *z* signal is to be maintained until the end of the cycle, its decoding AND gate can be dispensed with altogether.

The programmer of Fig. 2.5 might seem unduly complicated at first glance. However, its physical implementation can be quite simple provided the appropriate logic elements are used, as will be shown in Chapter 3. The programmer is especially advantageous where actuating valves with return springs are used; i.e., where sustained *z* signals are needed.

D. Reduced ring counter – Johnson code

The three full-ring-counter programmers described so far are simple to use, but are somewhat wasteful in the number of flip-flops required. At the cost of slightly more complicated circuits, it is possible to design programmers needing only one flip-flop for every two program steps. Such programmers will be called *reduced-ring-counter programmers*.

Since a reduced-ring-counter programmer provides two steps per flip-flop, every such programmer will accommodate an even number of steps. In cases where the program consists of an odd number of steps, the final step available from the programmer is not needed and is therefore 'wasted'. As soon as this last step is reached, the programmer is connected so that it immediately goes on to step No. 1. In other words, the last step is a 'dummy' step where nothing happens.

One possible programmer of the reduced-ring-counter type is based on the so-called Johnson code shown in Table 2.6. A programmer based on this code was described in [10], and is illustrated (eight stages) in Fig. 2.6. Note that the first half of the Johnson code is identical to the accumulating code. However, after all the flip-flops have been set, they are progressively reset at the rate of one flip-flop per step, rather than all simultaneously as in the accumulating code. Also note the similarity between the circuits in Figs. 2.5 and 2.6. Figure 2.6 merely has an additional AND gate inserted in the RESET line of each flip-flop.

This programmer requires only half the number of flip-flops as compared to a full-ring-counter programmer, but needs an additional AND gate per step for single-state decoding purposes. These extra AND gates needed may more than cancel the saving in flip-flops. However, this type of programmer has certain advantages nevertheless, and, as shown in Chapter 3, may be economical when certain switching elements are used.

Table 2.6 Johnson code

Step No.	y_1	y_2	y_3	y_4	Single-state decoding
1	1	0	0	0	$y_1 y_2'$
2	1	1	0	0	$y_2 y_3'$
3	1	1	1	0	$y_3 y_4'$
4	1	1	1	1	$y_1 y_4$
5	0	1	1	1	$y_1' y_2$
6	0	0	1	1	$y_2' y_3$
7	0	0	0	1	$y_3' y_4$
8	0	0	0	0	$y_1' y_4'$

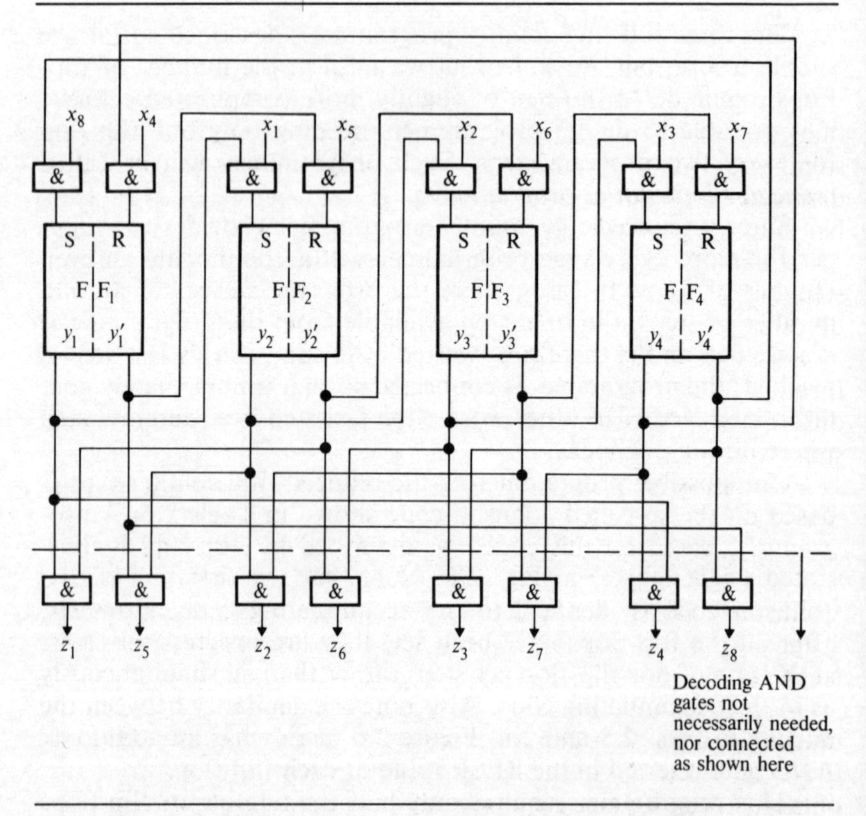

Fig. 2.6 Programmer based on Johnson code of Table 2.6 (eight stages)

If a given output signal z_i must be maintained for a number of m consecutive steps, multi-state decoding is used. This becomes especially convenient with the Johnson code. As shown in [10], all situations of multi-state decoding can be handled with a two-input AND or a two-input OR gate. The following rules can be stated, where m is the number of consecutive steps a z_i signal must be maintained, and n is the number of flip-flops in the programmer:

- For $m < n$, a two-input AND gate is required.
- For $m > n$, a two-input OR gate is required.
- For $m = n$, no gate is required. (The output signal is provided directly by one flip-flop output.)

The two gate inputs are obtained by selecting:

1. that flip-flop signal which distinguishes the first step where z_i is required from its preceding step;
2. that flip-flop signal which distinguishes the last step where z_i is required from its succeeding step.

For example, referring to Table 2.6, assume we wish to maintain an output signal z_i continuously from step No. 2 through step No. 4. Since $m = 3$ and $n = 4$, we have $m < n$. Step No. 2 is distinguished from step No. 1 by $y_2 = 1$, and step No. 4 from step No. 5 by $y_1 = 1$. We would thus use $z_2 = y_1 y_2$.

Assume now that we wish to maintain the output signal from step No. 2 through to step No. 5; i.e., $m = n = 4$. Here, no decoding gate is needed, and we would simply use $z_2 = y_2$.

Finally, for an output signal sustained from step No. 2 through to step No. 6, we have $m = 5$, so that $m > n$. Since $y_2 = 1$ differentiates between steps No. 1 and 2, and $y_3 = 1$ between steps No. 6 and 7, we would use an OR gate giving $z_2 = y_2 + y_3$.

E. Reduced ring counter – (1,2)/*n* code

Table 2.7 shows the (1,2)/n code, so called because alternately 1 and 2 out of n flip-flops are in the set condition in succeeding steps.

A programmer based on the (1,2)/n code must have a minimum of six steps. A few words may be in order to explain how the single-state decoding combinations listed in Table 2.7 are obtained. The y-decoding combination for any given step i applies equally to step i and step $i - 1$. Only the x signal in the full

Table 2.7 (1,2)/*n* code

Step No.	y_1	y_2	y_3	y_4	Single-state decoding
1	1	0	0	0	$y_1 y_2' x_8$
2	1	1	0	0	$y_1 y_4' x_1$
3	0	1	0	0	$y_2 y_3' x_2$
4	0	1	1	0	$y_2 y_1' x_3$
5	0	0	1	0	$y_3 y_4' x_4$
6	0	0	1	1	$y_3 y_2' x_5$
7	0	0	0	1	$y_4 y_1' x_6$
8	1	0	0	1	$y_4 y_3' x_7$

decoding expression differentiates between the two succeeding steps. The decoding expression for step *i* is found using one of the following expressions:

- For $i =$ even: $y_{i/2} \cdot y'_{(i/2)-1} \cdot x_{i-1}$
- For $i =$ odd: $y_{(i+1)/2} \cdot y'_{(i+3)/2} \cdot x_{i-1}$
- For one-before-last stage: $y_{(i+1)/2} \cdot y_1' \cdot x_{i-1}$

A programmer based on the (1,2)/*n* code was suggested in [8], and is shown in Fig. 2.7 (eight stages). In this programmer, every two program steps require only one flip-flop and two AND gates. The additional two AND gates required in the Johnson code programmer are not needed here, since the same AND gate serves for decoding purposes and also for supplying the SET and RESET signals at the same time. (This becomes possible due to the fact that the decoding combination of the *y* signals always overlaps two successive steps.) The drawbacks of this programmer are the somewhat more complicated connections between the elements, and the fact that AND gates with a fan-in of 3 are required. Therefore, this programmer is unsuitable for conventional spool valves, since these only have a fan-in of 2.

The problem of multi-state decoding for the (1,2)/*n* code programmer will now be discussed. As with 2/*n* code programmers, the output AND gates shown in Fig. 2.7 can never be dispensed with (even where additional decoding gates are used), since they are needed to supply the appropriate SET and RESET signals.

The programmer has the same limitation as that built according to the 2/*n* code. The input signals x_i must be sustained signals if we wish the corresponding output signals z_{i+1} to be sustained rather than pulses. We can repeat the discussion presented previously in connection with Fig. 2.4, though the solutions are not

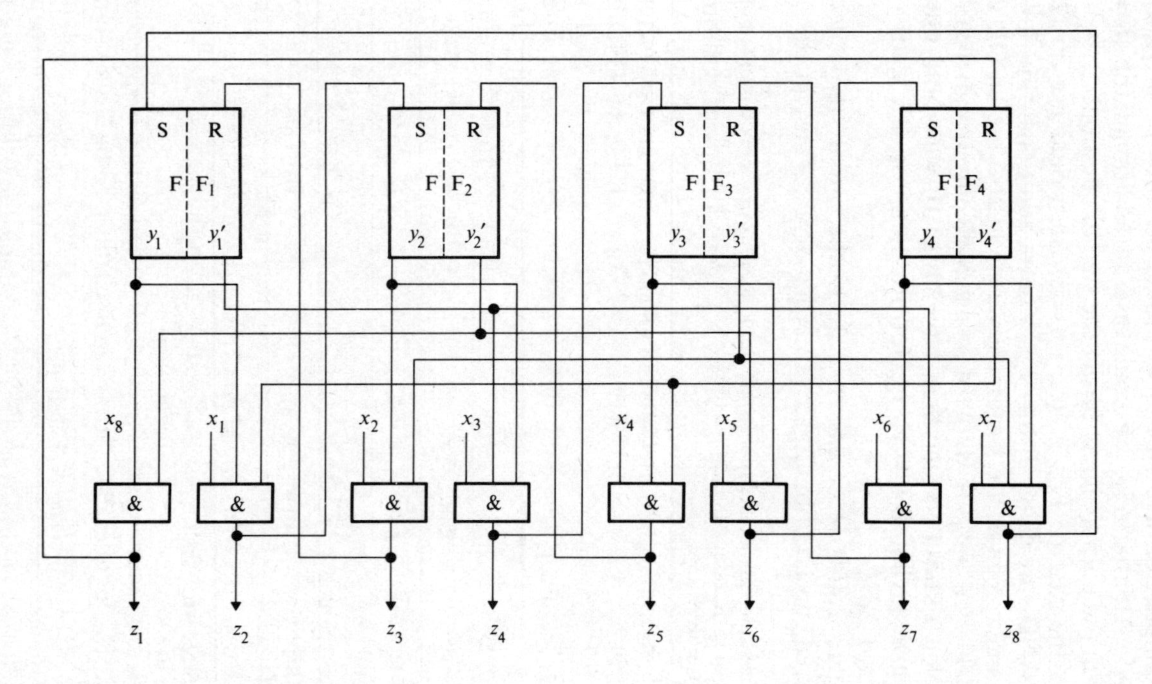

Fig. 2.7 Programmer based on (1,2)/*n* code of Table 2.7 (eight stages)

quite as simple any more. As before, there would be no problem in the case of Fig. 2.4(a). For Fig. 2.4(b), we would need an additional AND gate giving B+ $= y_1 y_2$. This solution, however, is only suitable for even-numbered steps, such as step No. 4 (Output $= y_2 y_3$), etc. As seen readily from Table 2.7, for odd-numbered steps we would need an AND gate with fan-in of 3. For example, for step No. 3, we would use the output function $y_1' y_2 y_3'$. Alternatively, we could use an additional flip-flop to maintain the output signal during the step in question.

For the case of Fig. 2.4(c), ($m = 2$), we would use an additional AND gate giving B+ $= y_2 y_3'$, or an additional flip-flop. If the x_1 and x_2 signals are sustained, we could also use an OR gate giving B+ $= z_2 + z_3$. If we need a sustained output signal during steps No. 3 and 4 (rather than No. 2 and 3), we would use an AND gate giving B+ $= y_1' y_2$, (or flip-flop, or OR gate with B+ $= z_3 + z_4$).

Figure 2.4(d) represents $m = 3$. If the initial step is even-numbered, no additional gate is needed at all. Thus, in the case

Table 2.8 Single- and multi-state decoding for (1,2)/n code programmer, showing necessary gate inputs to provide sustained output signal for m consecutive steps

m	Sustained output signal for steps No. i to $i+m-1$	i	Output function for sustained x signals	Output function for pulsed x signals
1	Step No. i	Even	z_i	$y_{i/2} \cdot y_{(i/2)+1}$
		Odd	z_i	$y'_{(i-1)/2} \cdot y_{(i+1)/2} \cdot y'_{(i+3)/2}$
2	Steps No. i to $i+1$	Even	$z_i + z_{i+1}$	$y_{(i/2)+1} \cdot y'_{(i/2)+2}$
		Odd	$z_i + z_{i+1}$	$y'_{(i-1)/2} \cdot y_{(i+1)/2}$
3	Steps No. i to $i+2$	Even	$y_{(i/2)+1}$	$y_{(i/2)/+1}$
		Odd	$z_i + z_{i+1} + z_{i+2}$	Flip-flop
4	Steps No. i to $i+3$		$z_i + z_{i+1} + z_{i+2} + z_{i+3}$ or flip-flop	Flip-flop
5	Steps No. i to $i+4$	Even	$y_{(i/2)+1} + y_{(i/2)+2}$	$y_{(i/2)+1} + y_{(i/2)+2}$
		Odd	Flip-flop	Flip-flop
≥ 6			Flip-flop	Flip-flop

of Fig. 2.4(d), we would simply connect $B+ = y_2$, since $y_2 = 1$ all during steps No. 2, 3 and 4. This becomes impossible if the initial step is odd-numbered. To avoid more complicated coding functions, it would be best here to fall back on the additional flip-flop. If the x signals are sustained, we can, of course, always use an OR gate with three inputs. The multi-state decoding rules for $(1,2)/n$ code programmers are summarized in Table 2.8.

CHAPTER 3

PROGRAMMER IMPLEMENTATION

The question of selecting one of the five programmer codes discussed in Chapter 2 as *the* optimum code does not permit a general answer. The matter depends, to a great extent, on the type of logic element that is to be used to implement the programmer, and also on the specific program. Certain codes turn out to be advantageous (or less so) in connection with specific types of logic elements.

For the purpose of the following discussion, we shall divide the various logic elements in general use into the following categories:

A. Pneumatic or hydraulic valves (either conventional or of miniature size), or equivalent moving-part fluidic elements.
B. Systems of moving-part fluidic elements providing the basic (NOT, AND, OR) switching functions.
C. Moving-part fluidic elements providing only a single complex switching function.
D. Moving-part fluidic or pure fluidic elements providing only the NOR function.
E. Wall-attachment fluidic elements.
F. Electronic logic gates.
G. Electromechanical relays.

In order to have some basis of comparison, we shall make use of a simple sample problem. Assume that we wish to actuate the two cylinders A and B shown in Fig. 2.3 according to the program

START, A+, B+, B−, A−, B+, B−, STOP

This program is displayed in the time–motion diagram of Fig. 3.1. The diagram also shows the required actuation of the pilot

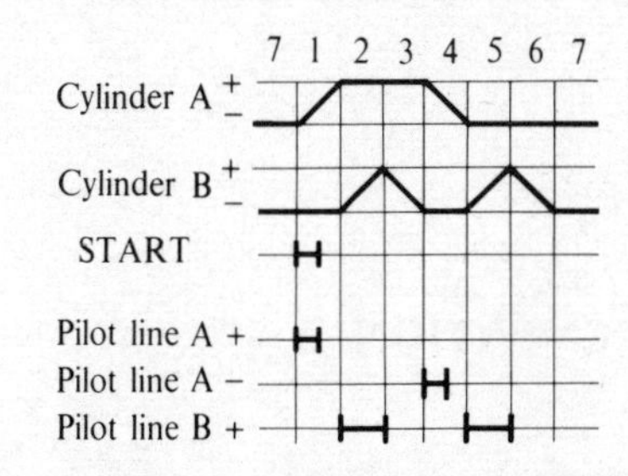

Fig. 3.1 Sample problem for system of Fig. 2.3

lines A+, A– and B+, for convenience in designing the various programmers. A program as simple as this one can, of course, easily be solved using various other methods. However, a simple program will serve better to illustrate the programmer design method.

The next five figures (Figs. 3.2 to 3.6) show the logic diagrams of the programmers needed to solve the above sample problem, using the five codes described respectively. Note that, for all cases, one OR gate is needed since the B+ pilot line has to be actuated twice during the cycle. (For other programs, the number of OR gates needed is liable to be different with different programmers, depending on the decoding used, as discussed in Chapter 2.) For the present program, multi-state decoding, as such, is not required, since the only sustained output signal (namely to pilot B+) is not caused by the appearance of an input pulse but rather by a sustained input signal ($a+$ or $a-$ respectively).

It should also be noted that step No. 7 in Fig. 3.1 does not require its own programmer stage. There are no required output signals during this step, nor are there any piston motions. This step therefore simply represents the STOP period between the end of the cycle (signal $b-$) and the beginning of the next cycle (START signal). In the programmers according to the $1/n$, $2/n$, Johnson and $(1,2)/n$ codes, this step No. 7 is accommodated by using an AND gate with the inputs $b-$ and START in the first stage, as shown in Figs. 3.2, 3.3, 3.5 and 3.6. In the accumulating code programmer (see Fig. 3.4), this step No. 7 corresponds to the state in which all flip-flop outputs $y_1 \ldots y_n$ are 0.

Comparing Figs. 3.2 to 3.6, we note that the total number of logic elements required for the different programmers is as follows:

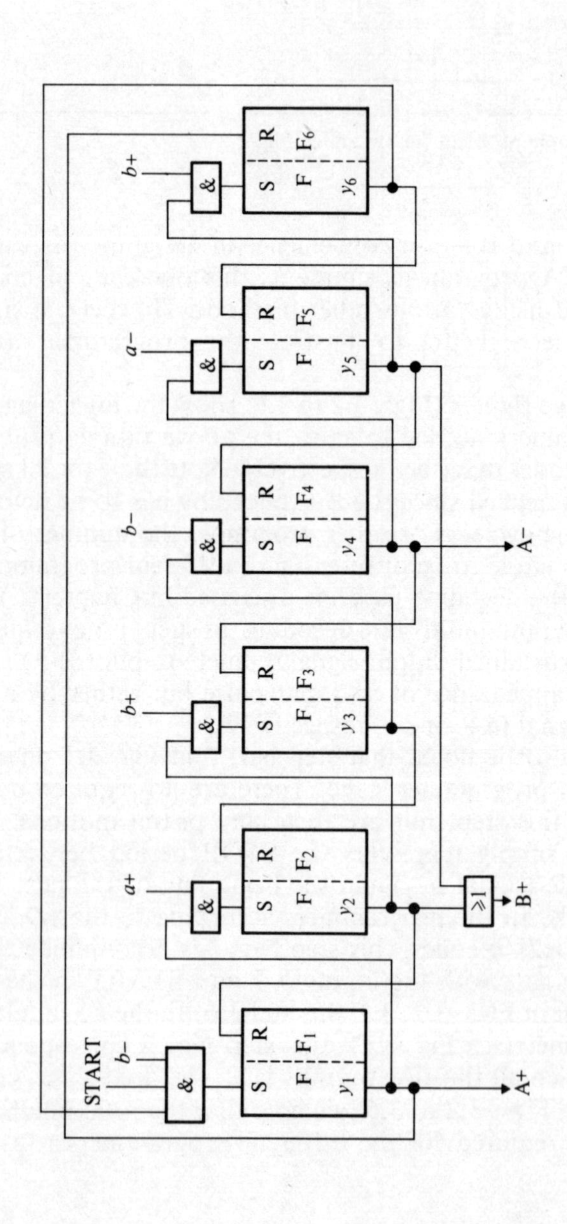

Fig. 3.2 1/*n* code programmer for sample problem of Fig. 3.1

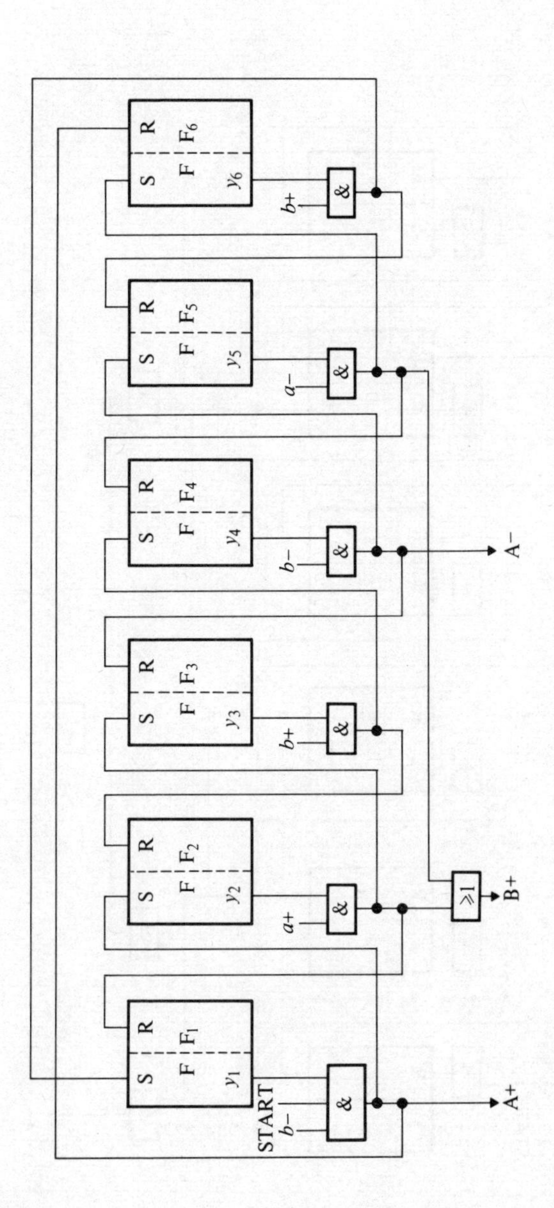

Fig. 3.3 2/*n* code programmer for sample problem of Fig. 3.1

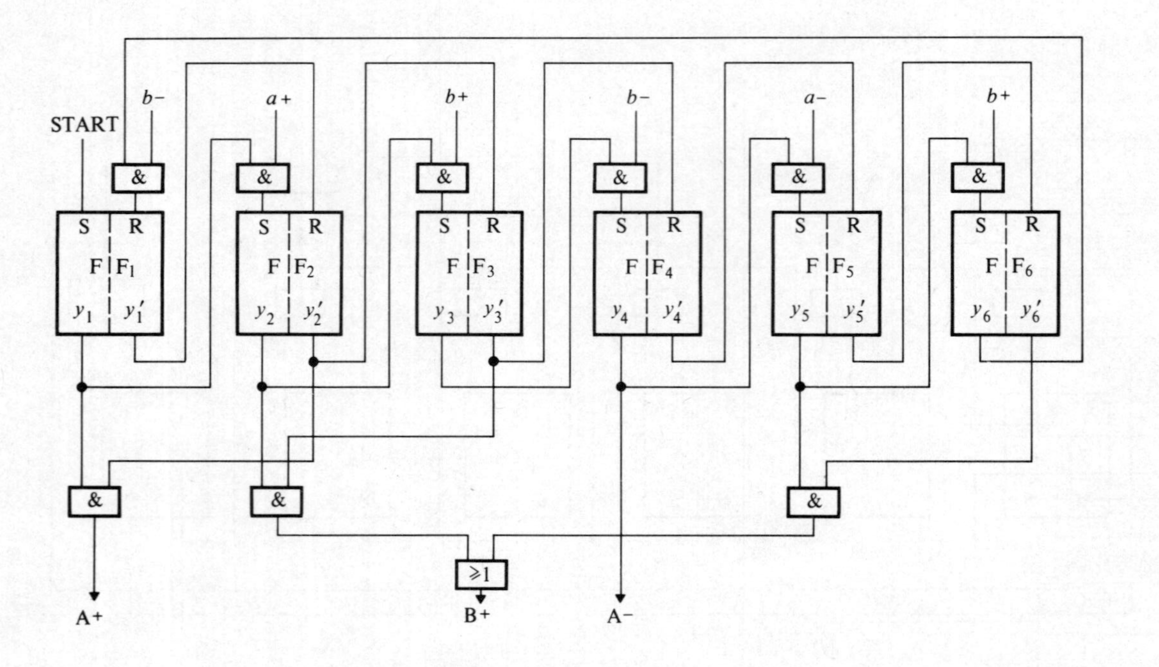

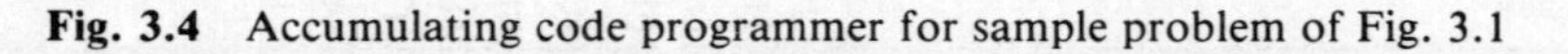

Fig. 3.4 Accumulating code programmer for sample problem of Fig. 3.1

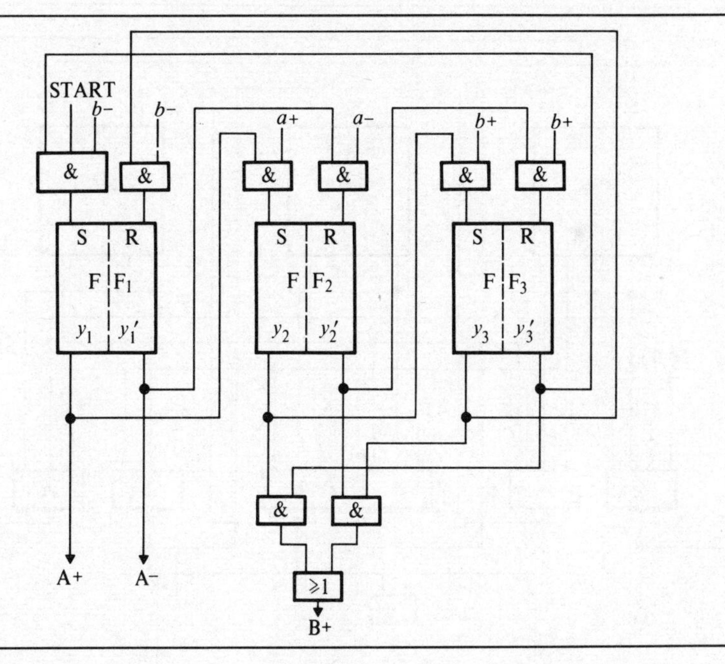

Fig. 3.5 Johnson code programmer for sample problem of Fig. 3.1

$1/n$ code	programmer:	13 elements
$2/n$ code	programmer:	13 elements
Accumulating code	programmer:	16 elements
Johnson code	programmer:	12 elements[1]
$(1,2)/n$ code	programmer:	10 elements

It would be premature, however, to draw general conclusions from the above tabulation. First of all, the above comparison only holds for the specific program of Fig. 3.1, which may by chance be more suitable for certain types of programmers and less so for others[1]. Secondly, the above count does not differentiate between flip-flops, AND gates and OR gates, although these different units do not, in general, have the same cost. Thirdly, the actual number of elements needed depends, to a very great extent, on the type of logic elements used, as will be shown in the following discussion.

A. Pneumatic or hydraulic valves, or equivalent elements

This category includes all normal valves of conventional size, or the miniature valves especially designed to serve in logic

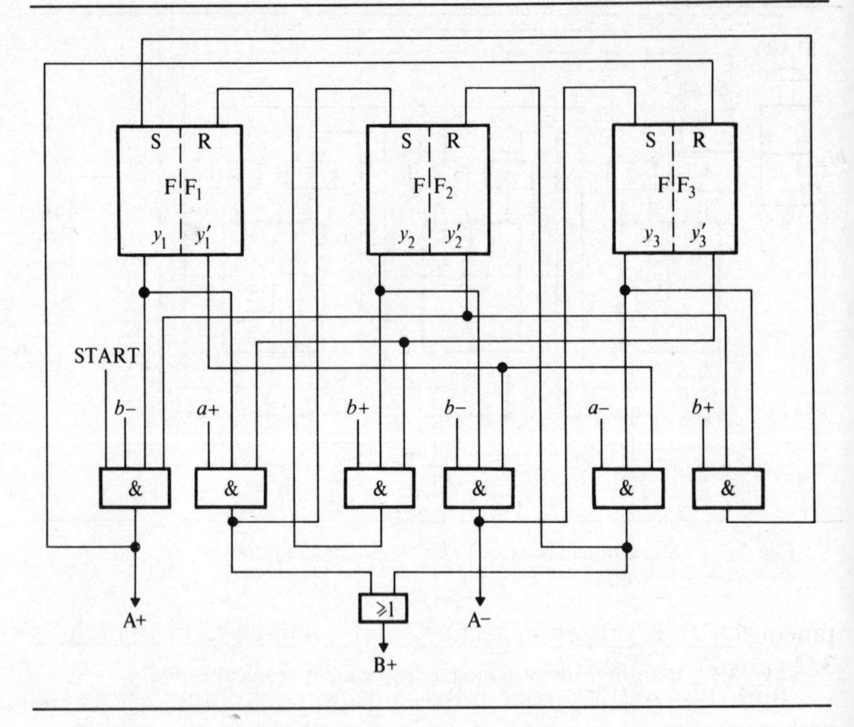

Fig. 3.6 (1,2)/*n* code programmer for sample problem of Fig. 3.1

circuits. We shall also include here those so-called moving-part fluidic elements which perform the same function as two-position three-way valves (here to be designated as 3/2 valves) without spring return but with double pilot lines, i.e. memory capacity. Some manufacturers only supply elements corresponding to five-way valves (5/2 valves), but these can also be used here (by plugging the second output connection), though they usually cost somewhat more. For examples of all of the above elements, see Category A in the Appendix.

3/2 valves with memory represent flip-flops with single outputs. Since the 1/*n* and 2/*n* code programmers do not utilize the complemented y' outputs of the flip-flops anyway, 3/2 valves are suitable only for these types of programmers. The remaining three codes do utilize the y' outputs, so that 5/2 valves would have to be employed there. (The accumulating code programmer circuit could be modified to get along without the y' outputs by using Inhibition rather than AND gates for generating the z outputs, and by actuating the RESET signals of all flip-flops simul-

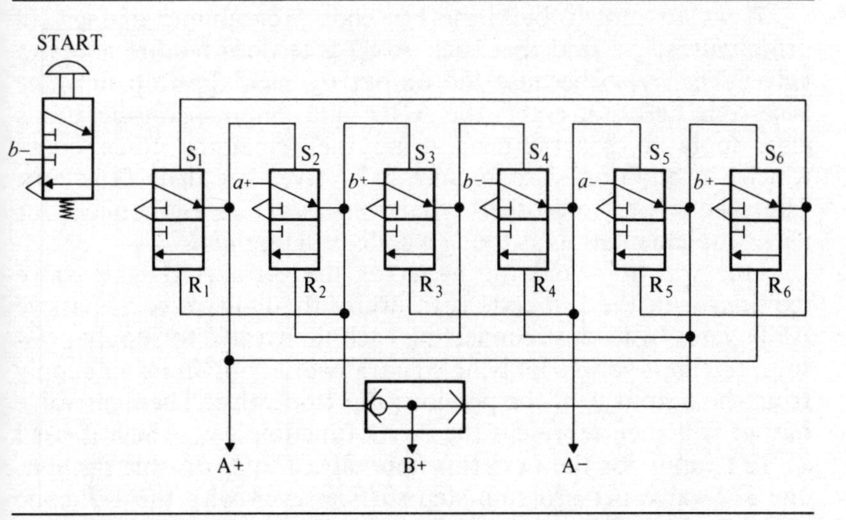

Fig. 3.7 Valve implementation of $2/n$ code programmer of Fig. 3.3

taneously, as discussed in section C of Chapter 2. In that case, 3/2 valves would suffice.)

Both $1/n$ and $2/n$ code programmers require one AND gate with each flip-flop. 2-input AND gates are ordinarily obtained using a 3/2 valve with single pilot line and spring return, the two inputs being the pilot pressure and supply pressure respectively. However, in the case of $2/n$ code programmers, the AND gate is connected to the flip-flop output, and we can therefore utilize the flip-flop valve to provide the AND function at the same time. This is shown in Fig. 3.7, which illustrates the 3/2 valve implementation of the $2/n$ code programmer of Fig. 3.3. A simple shuttle valve is used to provide the one OR gate needed.

As can be seen from the figure, the connections are extremely simple. Each valve output provides the SET signal for the succeeding valve, and the RESET signal for the preceding one. The supply pressures for the various valves come from the limit valves $a+$, $a-$, $b+$, $b-$, and all of these would be 3/2 valves with mechanical actuation and spring return.

Since no additional valves are needed for the AND gates, each program step requires only one valve (apart from OR gates, where needed). This makes the $2/n$ code programmer the most economical one and the favourite choice wherever the logic elements are pneumatic or hydraulic valves or other equivalent moving-part fluidic elements.

If we attempt to build the $1/n$ code programmer of Fig. 3.2 using valves, we find that each AND gate does require an extra valve. This is so because the output of each flip-flop must be accessible before it enters the AND gate. Similar considerations also apply to programmers using the remaining three codes which, in addition, also require 5/2 valves for their flip-flops. Thus, none of these other programmers are as economical for valve implementation as the $2/n$ code programmer.

One way to avoid the need for the extra AND-gate valve required with the $1/n$ code is to utilize the limit valve as passive AND gate. Instead of connecting each limit valve to supply pressure, the limit valve supplying signal x_i would obtain its air supply from the output y_i of the previous flip-flop valve. The limit valve output will then represent the AND function $x_i y_i$, which is used as SET input for the next flip-flop valve FF_{i+1}. In this fashion, one 3/2 valve per program step suffices even with the $1/n$ code programmer. This method, however, has two disadvantages: First, the amount of tubing required will increase appreciably if the limit valves are mounted at a distance from the programmer since, for each limit valve, we now require two tubing runs between valve and programmer. Second, we run into trouble with repeated program steps; i.e., where a given limit valve is actuated more than once during the cycle. For each such actuation, the limit valve is required to have different input and output connections. Since limit valves with multiple passages are not readily available, it becomes necessary to use a standard limit valve which, in turn, actuates several (depending on the number of required repeated events) additional 3/2 valves with pilot input and spring return.

It is interesting to note that one firm [18], markets ready-made programmer modules based on this principle. Each module contains three programmer stages, based on the $1/n$ code, with each stage requiring only a single 5/2 valve (apart from a shuttle-valve OR gate used for resetting the programmer stage externally). The logic diagram of one such module is shown in Fig. 3.8. The AND gates supplying the SET signals are the limit valves, which, as already explained, act as passive AND gates. The x inputs to these AND gates are not pressure inputs, but represent mechanical actuation of the limit valve by the piston rod. The AND gates shown at each flip-flop output do not require additional valves, but are obtained by connecting a normally-closed flow passage of valve i in series with a normally-open flow passage of the previous valve $i-1$. In this way, the z outputs are

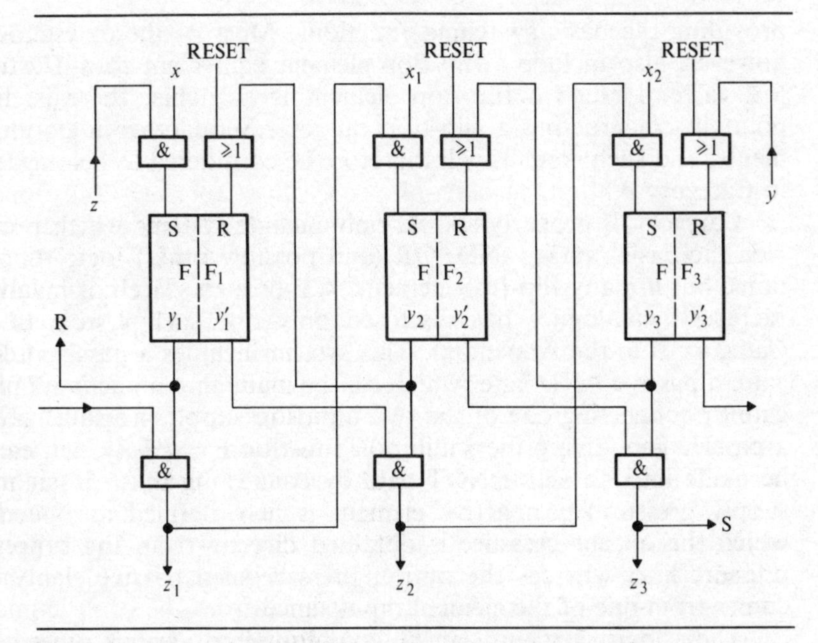

Fig. 3.8 Three-stage programmer module based on $1/n$ code [18]

operated in a 'break-before-make' mode; i.e., the previous flip-flop must first be reset, before the next z output is turned on. The manufacturer claims that this provides an additional safety interlock, which will cause the program to come to a halt in case a given valve gets stuck in its SET position. However, operating the z outputs in a 'break-before-make' manner might be problematic where continuous z signals are required. The manufacturer also markets 3/2 valves with single pilot actuation and spring return, having a triple set of independent flow passages. Each such valve will be able to accommodate three repetitions of a given program step during the cycle.

It must be pointed out that the programmer module described above will cost somewhat more than a $2/n$ code programmer put together by the user himself out of standard off-the-shelf 3/2 valves. On the other hand, the ready-made programmer module eliminates some of the unsightly tubing, and will undoubtedly be more compact and convenient.

B. Systems of moving-part fluidic elements providing the basic (NOT, AND, OR) switching functions

A number of manufacturers market systems of logic elements

providing the basic switching functions. Most of these systems, however, also include a flip-flop element equivalent to a 3/2 or 5/2 valve. If such a flip-flop element is available, there is no point in constructing a flip-flop out of several other logic elements, and such systems will therefore be considered as belonging to Category A.

Category B properly should only include systems which provide the basic NOT, AND, OR (and possibly other) logic functions, but *not* any flip-flop element. A thorough search of manufacturers' catalogues has disclosed only one such system (see Category B in the Appendix). This system includes a passive OR gate, a passive AND gate (which can be made into an active YES gate by connecting one of the two inputs to supply pressure) and a passive gate giving the Inhibition function $F = a'b$ (which can be made into an active NOT gate by connecting the b input to supply pressure). An active element is here defined as one in which the output pressure is obtained directly from the supply pressure line, whereas the output pressure of a passive element comes from one of the element input signals.

These logic elements can be combined into various types of feedback circuits in order to obtain flip-flops, and two of the most common ones are shown in Fig. 3.9. The first, Fig. 3.9(a), has the 'RESET dominating' characteristic; i.e., if both S and R signals should be 1 simultaneously, the flip-flop output y will be 0. The second, Fig. 3.9(b), is 'SET dominating', giving $y = 1$ for $S = R = 1$. Note that both flip-flops have single outputs; i.e., the complemented y' output is not available.

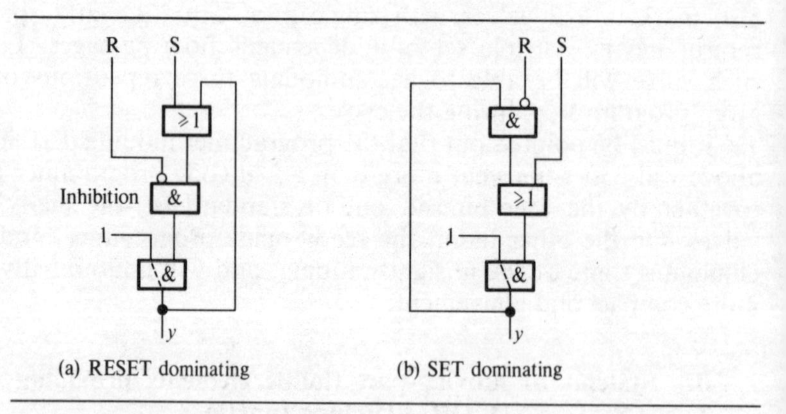

Fig. 3.9 Feedback circuits providing single-output flip-flops (a) RESET dominating (b) SET dominating

In both of the above flip-flop circuits, the bottom YES gate would not be needed if one of the other two gates were an active element. Since we are here dealing with passive elements, the YES gate is needed to provide an active pressure source for the feedback loop. Without such an active source, the feedback pressure would gradually drop because of small leakages, so that the flip-flop memory would eventually collapse. The dashed lines drawn through the AND gates in Fig. 3.9 indicate that the actual output pressure is obtained from the supply-pressure source, and not from the second signal input connected to the previous gate.

It is possible to construct $1/n$ or $2/n$ code programmers using the above elements, but such programmers would require four elements per stage, namely three for the flip-flop and one for the AND gate associated with each stage (apart from additional OR gates or other decoding gates, where required). The Johnson code turns out to be the most economical choice for Category B elements. However, the accumulating code does not lag far behind, and would most probably be preferred because of the greater simplicity of the circuit connections. Figure 3.10 shows the logic-gate implementation of the accumulating code programmer of Fig. 3.4.

The circuit of Fig. 3.10 seems to be quite different from Fig.

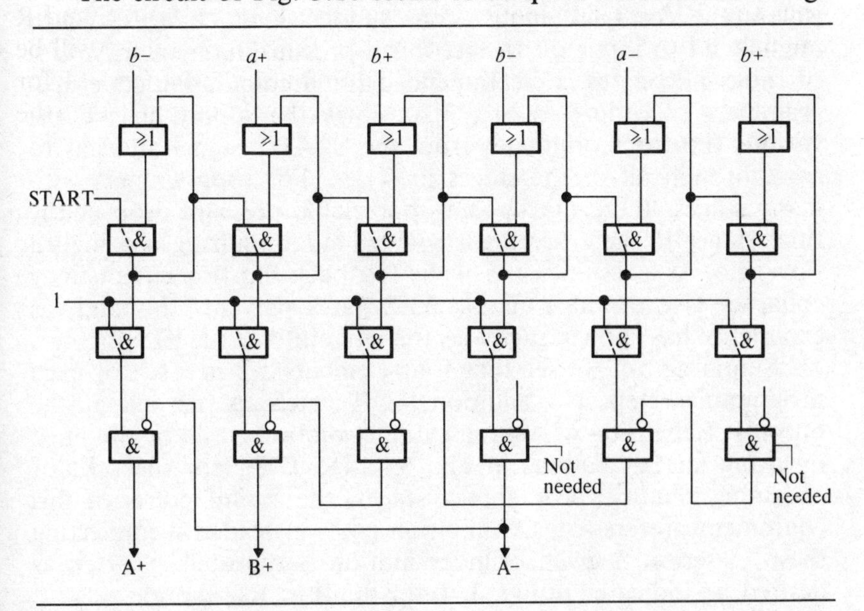

Fig. 3.10 Logic gate implementation of accumulating code programmer of Fig. 3.4

3.4. Nevertheless, careful comparison will show that all the functions indicated in Fig. 3.4 are performed by the circuit of Fig. 3.10. The AND functions shown at the input of each flip-flop in Fig. 3.4 are provided by each upper AND gate in Fig. 3.10. Once a given flip-flop is set, its upper AND gate provides a 1-input to the upper AND gate of the next flip-flop, thus preparing it to receive the next input signal. There are no RESET signals as such, nor are they needed for the accumulating code. At the end of the cycle, all the flip-flops will be set, and the programmer will automatically stand still because the Inhibition gate connected to A+ will have a 0 output. To get the next cycle started, it is first necessary to interrupt the START signal. This withdraws the air supply from the upper AND gate of the first flip-flop. This, in turn, withdraws the 1-signal from the next upper AND gate, so that, one by one, all the flip-flop circuits collapse. Renewal of the START signal causes the next cycle to commence.

Unlike in the circuit of Fig. 3.9(a), it is now possible to take the feedback signal from the output of each upper AND gate, since these gates are now, in practice, active ones. This is so because supply air is supplied continuously through the upper AND gate of the first stage. Nevertheless, it is advisable to add an additional YES gate at the output of each stage, in order to isolate the feedback loops from sudden load changes, which might tend to interrupt the feedback pressure momentarily. Use of these YES gates is recommended for accumulating code programmers according to Fig. 3.10, since the supply air for the various flip-flops originates from the START signal and has to pass through all of the stages in series. For programmers with many stages, there may be an appreciable pressure drop at the final stages if these are loaded with loads requiring too high a flow rate. As a result, some of the feedback flip-flop circuits may collapse. Use of the isolating YES gates prevents this and, as experience has shown, increases the reliability of the circuits.

Summing up, we see that four elements are needed for each programmer step. No additional OR gates are needed at the outputs. Repetition of a given cylinder motion could be obtained by using an OR gate as in Fig. 3.4. However, for the sake of obtaining identical programmer stages, the manufacturer of this equipment prefers to use Inhibition gates instead. By connecting these in series, a given cylinder motion is repeated as often as desired, as indicated in Fig. 3.10 for the B+, B− motion.

It is interesting to note that the manufacturer of this equipment supplies ready-made programmer units containing up to 16 stages, each stage having one OR, two AND and one Inhibition gates, with all internal connections already made. The user need only connect the x_i input signals and the z_i outputs to the programmer, and connect the appropriate Inhibition gates. Apart from saving assembly costs, the unit has the advantage of compactness. Outside indicators show the state of each flip-flop, so that one can tell at any time what step the programmer is at.

C. Moving-part fluidic elements providing only a single complex switching function

These are represented by the elements listed under Category C of the Appendix. These elements are also called 'multi-function elements', since each element can be made to give all the basic logic functions. The discussion is somewhat complicated by the fact that each one of these elements provides a different switching function, and must therefore be discussed separately.

Considering the first element listed under Category C, we are dealing with a four-input element giving the output function $F = AC + B'C + A'D + BD$. By setting one or more of these four inputs to be either 1 or 0, all the basic logic functions can be obtained. For example, setting $C = 1$ and $D = 0$, we get the Implication function $F = A + B'$. (While the OR function is also obtainable from this element, it is usually obtained at less cost by means of a special inexpensive OR element.) A flip-flop can be obtained by using one of the two methods indicated in Fig. 3.9, or by connecting two Implication gates ($F = A + B'$) in a feedback circuit as shown in Fig. 3.11, which has the advantage that both outputs y and y' become available.

The manufacturer of these elements supplies ready-made programmer units containing three or six stages, and these are

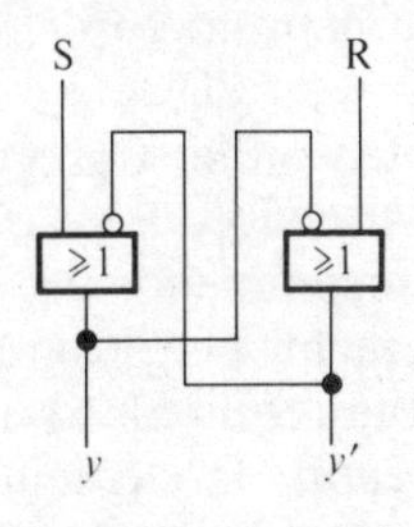

Fig. 3.11 R–S flip-flop consisting of two Implication gates

identical to the accumulating code programmer illustrated in Fig. 3.10. Thus, 24 elements would be needed to solve the sample problem of Fig. 3.1 (or only 22 elements, if we do not insist on identical stages and thus cancel the unnecessary Inhibition gates in stages No. 4 and 6).

By comparison, the $1/n$ and $2/n$ code programmers of Figs. 3.2 and 3.3 respectively would require only 19 elements, using flip-flops according to Fig. 3.11 (i.e., 6 AND gates, 1 OR gate and 12 elements for the 6 flip-flops, for a total of 19 elements). This calculation assumes, just as in similar comparisons to be made later, that the START button is inserted into the $b-$ signal line, just as in Fig. 3.7, so as to avoid the need for an AND gate with fan-in = 3.

The $2/n$ code programmer of Fig. 3.3, while also requiring 19 elements, has the slight advantage that, if the dead-ended input of each AND gate is connected to the flip-flop output y_i, the flip-flop would be completely isolated from all loads, without the need of additional YES gates. Use of such YES gates in $1/n$ code programmers is recommended by some manufacturers for greater reliability, even though they are probably not really necessary.

The Johnson code programmer of Fig. 3.5 would require 6 elements for the 3 flip-flops, 8 AND gates, and 1 OR gate, for a total count of 15 elements. Finally, the $(1,2)/n$ code programmer of Fig. 3.6 requires 6 elements for the 3 flip-flops, 12 AND gates (since 3-input AND gates are needed, and these require an additional element each) and 1 OR gate, for a total of 19 elements. Since the $(1,2)/n$ code programmer has more complicated connections than the $1/n$ or $2/n$ code programmers, there is no advantage in using it here.

Summarizing, we conclude that the Johnson code programmer is the most economical one (15 elements) for the sample problem used. The $1/n$ and $2/n$ code programmers do not lag far behind (19 elements). It is interesting to note that the accumulating code programmer, which is the one used by the manufacturer, comes out worst in this particular comparison.

We turn now to the second element listed under Category C, a four-input element having the output function $F = AB'C + D(A' + B)$. The manufacturer of this unit recommends use of the $1/n$ code programmer [5]. Basically, two elements per programmer stage should here be sufficient to supply the required AND gate and flip-flop. However, just as with Category B elements, the manufacturer himself uses a third element as YES gate at the output of each stage in order to isolate the flip-flop from the

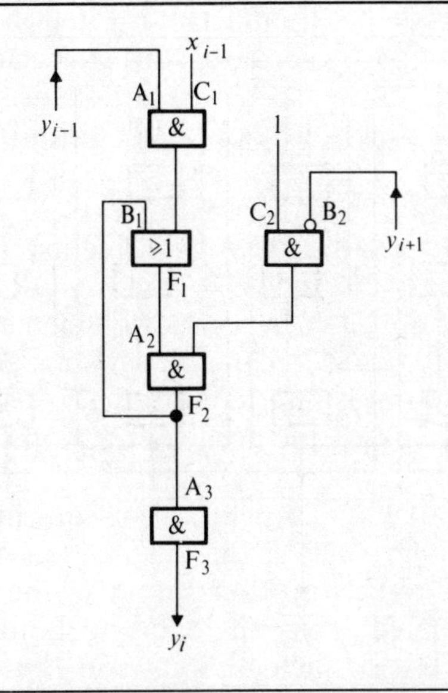

Fig. 3.12 Stage of $1/n$ code programmer, implemented with three elements having switching function $F = AB'C + D(A' + B)$

load. The recommended programmer stage uses flip-flops somewhat similar to that shown in Fig. 3.9(a), and is illustrated in Fig. 3.12, where the three elements of each stage give the following output functions F_1, F_2 and F_3 respectively:

First element: $B_1 = D_1$ giving $F_1 = B_1 + A_1C_1$
Second element: $D_2 = 0$ giving $F_2 = A_2B_2'C_2$
Third element: $B_3 = D_3 = 0$; $C_3 = 1$ giving $F_3 = A_3$

Implementing the $1/n$ code programmer of Fig. 3.2 by means of such programmer stages would require 19 elements ($6 \times 3 = 18$ elements for the 6 stages, plus 1 element for the OR gate). If we do not use the isolating YES gates at the output of each stage, 13 elements will suffice.

A $2/n$ code programmer could be built out of two elements per stage as follows:

First element: $C_1 = 1$ giving $F_1 = AB' + D(A' + B) =$
$D_1 + A_1B_1'$
Second element: $B_2 = D_2 = 0$ giving $F_2 = A_2C_2$

Using the first element (which is an active one) to build a flip-flop similar to Fig. 3.9(b), and the second element to provide the AND function and at the same time to isolate the flip-flop

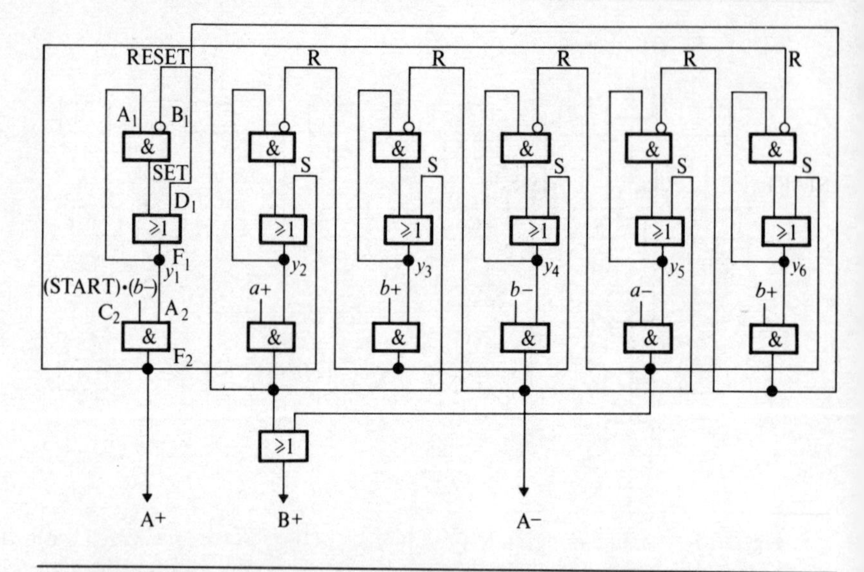

Fig. 3.13 2/*n* code programmer of Fig. 3.3 implemented with 13 elements having switching function $F = AB'C + D(A' + B)$

from the load (since the flip-flop output is connected to the dead-ended A input of the AND gate), we obtain the implementation of the 2/*n* code programmer of Fig. 3.3 shown in Fig. 3.13. Altogether only 13 elements are needed.

Johnson and (1,2)/*n* code programmers would not be suitable here, since the complemented flip-flop outputs y' are not available. Thus, only the accumulating code remains to be discussed.

An accumulating code programmer would seem to need additional elements for the output Inhibition gates. However, the output Inhibition function $F = AB'$ is obtained here by means of an active element (setting $C = 1$ and $D = 0$), so that these Inhibition gates can take the place of the output YES gates. The programmer would therefore look as in Fig. 3.10, but without the lower row of AND gates, so that only 18 elements would be needed.

Even greater economy can be obtained if we are willing to use the complemented input signals x_i' rather than x_i. These can easily be obtained with most limit valves by interchanging the air-supply and vent connections, so that the valves will pass air when not actuated, and vent the air when actuated. Once these x_i' signals are available, we can obtain the necessary flip-flop stage by means of a single element, apart from the second element

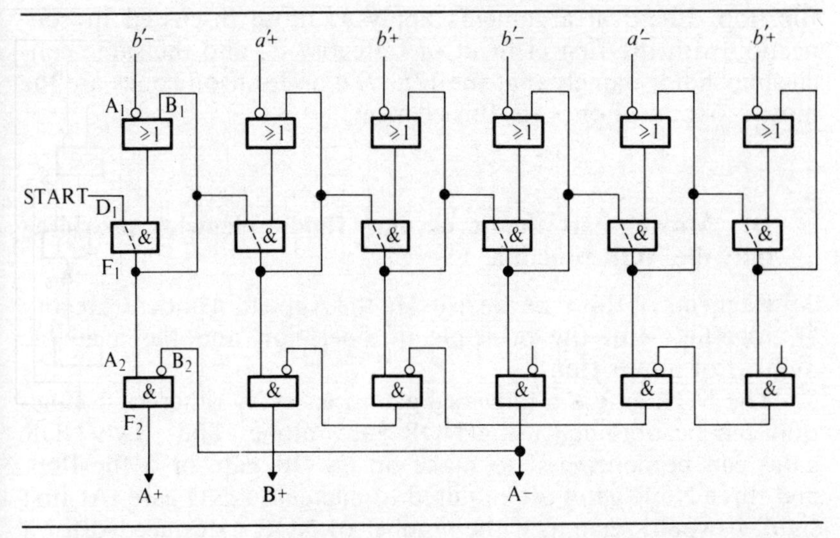

Fig. 3.14 Accumulating code programmer of Fig. 3.4 implemented with 12 elements having switching function $F = AB'C + D(A' + B)$, using complemented input variables

needed for the output Inhibition function. The two elements would be utilized as follows:

First element: $C_1 = 0$ giving $F_1 = D_1(A_1' + B_1)$
Second element: $C_2 = 1$; $D_2 = 0$ giving $F_2 = A_2B_2'$

The resulting implementation of the accumulating code programmer of Fig. 3.4 is shown in Fig. 3.14. Altogether, only 12 elements are needed.

Summarizing, the accumulating code programmer is the most economical one for this type of element, provided it is possible to use complemented input signals x_i' rather than x_i. If this is not possible, the $2/n$ code programmer should be used. The $1/n$ code programmer can only compete from the standpoint of element economy if we do not insist on the isolating YES gates at the output of each stage. It is interesting to note that, just as with the previous element discussed, the preferred codes are not those used by the manufacturer.

We now come to the third element listed under Category C, a three-input element having the output function $F = AB + A'C$. Setting $C = 1$, the Implication function $F = AB + A' = A' + B$ is obtained. As was shown in Fig. 3.11, a flip-flop can be constructed out of two Implication gates, or flip-flops can be built according to Fig. 3.9. In either case, we need two elements per flip-flop. Identical arguments apply as those discussed in connection with the first element of Category C, and the same conclusions hold, namely that the $1/n$, $2/n$ or Johnson codes are the most economical ones for this element.

D. Moving-part fluidic or pure fluidic elements providing only the NOR function

Elements of this type are listed in the Appendix under Category D, together with the principle of operation and the available number of inputs (fan-in).

The NOR gate is a universal gate; i.e., every other logic function can be obtained using NOR gates alone. Thus, two NOR gates can be connected to make up an OR gate or a flip-flop, and three NOR gates are required to make an AND gate. At first sight, it would seem as if the number of NOR gates needed for a given circuit will be very large. However, by proper design, the number of NOR gates needed can be greatly reduced. Also, the relatively low price of some of the available NOR gates partially makes up for the additional gates needed.

Since a NOR gate with the inputs A and B gives an output $T = A'B'$, we see that the NOR gate can take the place of an AND gate, provided the two gate inputs are the complements of the inputs A and B of the required AND function. In other words, if we connect the inputs C' and D' to a NOR gate, the gate output will be $T = CD$, which represents the AND function. We can thus obtain what amounts to an AND gate using only one instead of three NOR gates.

In the case of programmers, this means that we must interchange the y and y' outputs of the various flip-flops, and must also use the x_i' rather than the x_i signals from the various limit valves. As already mentioned, these x_i' signals are obtained by interchanging the air-supply and vent connections, so that the limit valve will pass air when not actuated, and vent the air when actuated.

The number of NOR gates required for the various types of programmers is tabulated below.

If NOR gates with a fan-in of 3 are available, the $(1,2)/n$ code programmer is by far the most economical one. The NOR gate implementation of the $(1,2)/n$ code programmer of Fig. 3.6 is shown in Fig. 3.15. If a programmer of simpler construction is preferred, the $1/n$ or $2/n$ code programmers should be used.

$1/n$ code	programmer (according to Fig. 3.2):	20 NOR gates
$2/n$ code	programmer (according to Fig. 3.3):	20 NOR gates
Accumulating code	programmer (according to Fig. 3.4):	23 NOR gates
Johnson code	programmer (according to Fig. 3.5):	16 NOR gates
$(1,2)/n$ code	programmer (according to Fig. 3.6):	14 NOR gates (Fan-in = 3)

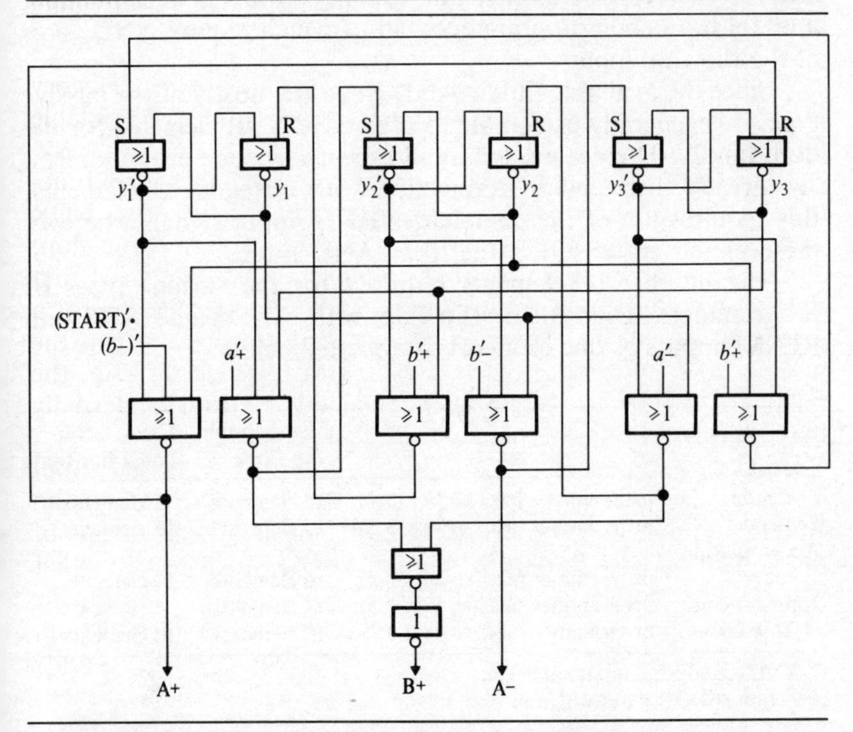

Fig. 3.15 $(1,2)/n$ code programmer of Fig. 3.6 implemented with 14 NOR gates (Fan-in = 3)

If the available NOR gates have a fan-in of only 2, then two additional gates would be required per flip-flop to provide the OR inputs necessary for resetting the programmer after a power interruption (see Chapter 4, Section A). Since the accumulating code is the only one that does not require such OR inputs, it would become the most economical choice for NOR gates with fan-in of 2.

E. Wall-attachment fluidic elements

Elements of this type are listed in the Appendix under Category E.

Both flip-flops and OR/NOR functions can each be obtained with one single wall-attachment fluidic element. Some manufacturers supply standard flip-flop elements with passive AND gates already built into the SET and RESET input lines. These are, of course, much cheaper than individual AND gates and flip-flops purchased separately and then connected by the user, and can therefore be used advantageously with $1/n$, accumulating and Johnson code programmers, all of which require AND gates at the flip-flop inputs.

Since the available fluidic AND gates are mostly of the passive type, it is generally preferable to use active NOR elements for the decoding AND gates needed in all except $1/n$ code programmers. As already discussed in connection with Category D elements, this requires use of the complemented x_i' input signals wherever these signals enter NOR rather than AND gates.

The number of elements required for the various types of programmers (counting a flip-flop with AND-gated SET and RESET inputs as one element) is tabulated below:

		Using simple flip-flops	Using AND-gated flip-flops
$1/n$ code	programmer (acc. to Fig. 3.2):	13 elements	7 elements †
$2/n$ code	programmer (acc. to Fig. 3.3):	13 elements	13 elements
Accumulating code	programmer (acc. to Fig. 3.4):	16 elements	10 elements †
Johnson code	programmer (acc. to Fig. 3.5):	12 elements	6 elements
$(1,2)/n$ code	programmer (acc. to Fig. 3.6):	10 elements ‡	10 elements ‡

† AND-gated SET input sufficient
‡ Requires NOR gates with fan-in = 3

The tabulation shows that, if simple flip-flops are used, the $(1,2)/n$ code programmer is the most economical, provided 3-input NOR elements are available. However, the advantage of the $(1,2)/n$ code programmer is not that pronounced here, and it may be preferable to use the $1/n$ or $2/n$ code programmer for the sake of having simpler connections.

If flip-flop elements with AND-gated inputs are available, we would gain by using the $1/n$ code programmer (AND-gated SET inputs are sufficient here), or the Johnson code programmer (AND gates required in both SET and RESET inputs). The use of these AND-gated flip-flops is, however, somewhat problematical. Passive AND gates of this type work reliably only if the two input pressures reaching the gate are of similar magnitude.

As used in the circuits of the $1/n$ and Johnson code programmers, one input signal of the passive AND gate is always supplied by a flip-flop output, and will therefore be at a fairly low pressure level (approximately 25–40% of the fluidic element supply pressure). The second input signal, on the other hand, is the x_i signal which may be of very high or very low pressure level, depending on the type of valve or position sensor supplying the signal. An additional fluidic element could be inserted between sensor and programmer to supply an x_i signal having the correct pressure level, but the benefit of element economy associated with the use of AND-gated flip-flops would then be lost.

F. Electronic logic gates

The cost of electronic logic gates (especially those made of integrated circuits) has come down to such an extent that the necessary peripheral equipment, packaging and connections often cost more than the gates themselves. For this reason, gate economy assumes secondary importance here, while ease of system design may be the primary consideration.

The design of the control system usually becomes easiest with $1/n$ code programmers, for the following reasons:

(a) No need for sustained x_i signals (see Ch. 2, sect. B).
(b) Easier to adapt for multi-path programs (see Ch. 5).
(c) The complemented output signals z_i' are automatically available from the second output of each flip-flop (provided double-output flip-flops are used), which may facilitate the design of output logic networks.

For the above reasons, most manufacturers of electronic programmers supply circuit cards containing assembled $1/n$ code programmer stages. Often, the flip-flop SET and RESET inputs have multiple-input AND and OR gates respectively, which especially facilitates the design of multiple-path programs.

In the case where the user wishes to put together his own programmer using individually packaged gates, the reasoning is somewhat different, since gate economy is now a factor. Assuming that a flip-flop costs approximately the same as two gates, the comparison between the various codes leads to similar results as for Category D (NOR) elements. Though the $(1,2)/n$ and Johnson code programmers need the least number of elements, the $1/n$ and $2/n$ code programmers are probably preferable here because of their inherent simplicity. The $2/n$ code programmer has the advantage that the AND gates connected after each flip-flop automatically isolate the flip-flop circuits from the load.

G. Electromechanical relays

Each relay can be made to serve as one flip-flop. Relays with a fairly large number of contacts are readily available, and the cost of the relay does not rise too sharply with the number of contacts. The various logic functions are obtained simply by connecting the relay or switch contacts in series (AND gate) or in parallel (OR gate). Thus, no additional relays are needed for these gates, and the number of relays needed will exactly equal the number of required flip-flops. The most economical programmer will therefore use either the Johnson or $(1,2)/n$ code requiring one relay for each two program steps. The relay implementation of the $(1,2)/n$ code programmer of Fig. 3.6 is shown in Fig. 3.16, which is arranged in the form of a ladder diagram, as is customary for relay circuits.

In studying Fig. 3.16, two facts become obvious:

1. The number of limit-switch contacts required for most programmers will be fairly large. While, as pointed out above, additional relay contacts are readily available, the same cannot be said of limit-switch contacts. Most standard off-the-shelf limit switches have only one pair of contacts or, at most, single-pole double-throw (i.e., change-over) contacts. If more contacts are required, non-standard switches must be purchased, and these are liable to be expensive. We could, of course, always use a switch with a single pair of contacts to actuate a relay coil, and this relay will then supply the additional contacts we need, but this will increase the number of relays required.
2. The number of contacts and connections is very large, so that the programmer has lost the advantage of simplicity. This is due to the inherent difference between logic-gate outputs and relay or switch contacts. When using valves or ordinary logic gates, each binary variable can be connected to as many inputs as are required (subject to the allowable fan-out of the element). As a result, the programmer can be built up out of standard modules, as demonstrated in the previous sections. The finished programmer then assumes the form of a pre-

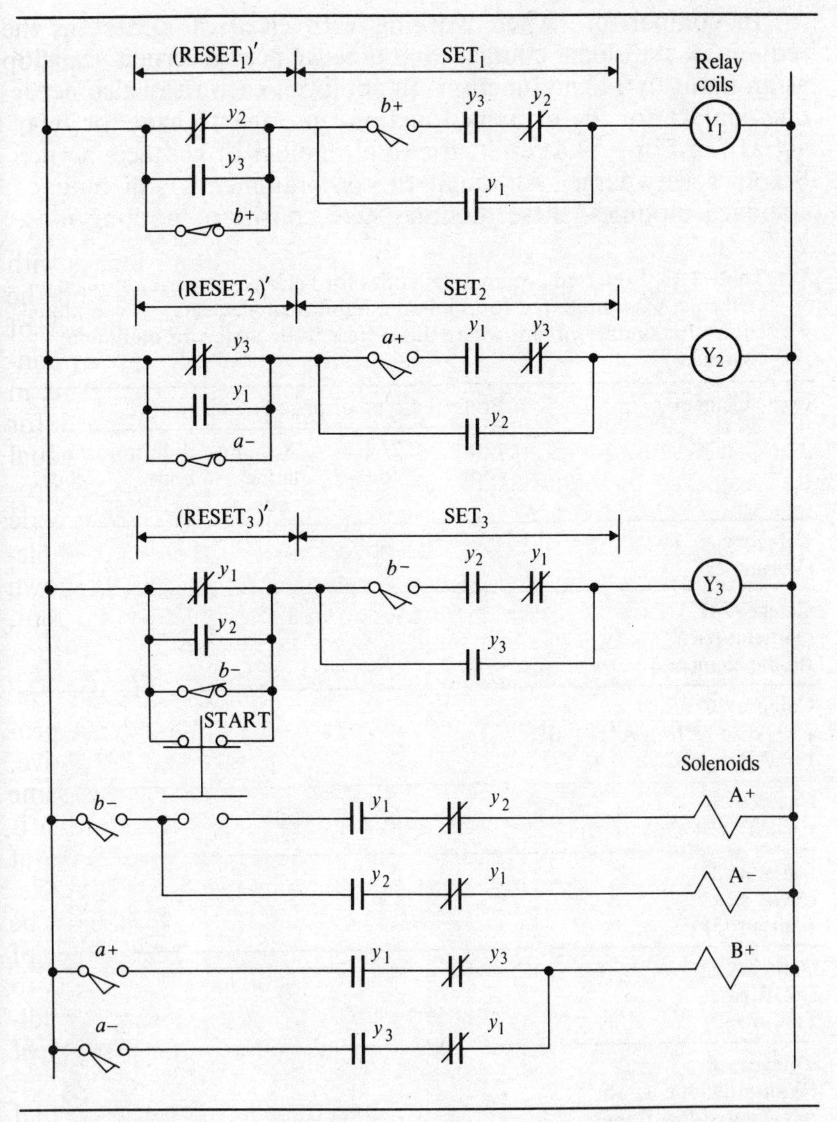

Fig. 3.16 Relay implementation of (1,2)/n code programmer of Fig. 3.6

fabricated 'black box' with all the internal SET and RESET connections already made, and with a set of input connections for the input variables $x_1 \ldots x_n$, and a set of output connections for the output variables $z_1 \ldots z_n$. All that remains to be done is to connect these input and output variables to appropriate points on the programmer.

In comparison, when working with electrical contacts, we require an additional contact each time we need a certain variable as an input to a logic function. In addition, each relay also needs one contact for the memory function, in order to have the relay act as flip-flop. As a result, the total number of contacts needed becomes very large. Although the programmer is still built of standard modules, these modules here consist of nothing more

Table 3.1 Preferred programmer codes for various categories of logic elements. ⊗ = preferred for minimum number of elements; ○ = optional choice for simpler circuit, where this differs from choice for minimum elements

Logic element	Preferred type of programmer				
	$1/n$ code	$2/n$ code	Accumulating code	Johnson code	$(1,2)/n$ code
Category A (Valves)		⊗			
Category B (Moving-part fluidic elements)			○	⊗	
Category C F = AC + B′C + A′D + BD F = AB + A′C	○	○		⊗	
Category C F = AB′C + D(A′ + B)		⊗	⊗		
Category D (NOR gates Fan-in = 3)	○	○			⊗
Category D (NOR gates Fan-in = 2)			⊗		
Category E (Wall-attachment, no AND-gated flip-flops)	○	○			⊗
Category E (Wall-attachment, with AND-gated flip-flops	⊗			⊗	
Category F (Electronic gates)	○	○			⊗
Category G (Relays)				⊗	⊗

than a relay coil and its memory contact. The SET and RESET connections are missing, and so many additional contacts and connections are required that we have almost completely lost the desirable characteristics of the ring-counter programmers, namely simplicity and flexibility.

We must therefore come to the reluctant conclusion that electrical relays are not very suitable for use in ring-counter programmers.

H. Conclusions

We shall now attempt to summarize the results of this chapter. For most categories of logic elements, it is difficult to point to one type of programmer as the optimal one for all cases. This is because the final choice depends on the type of program required, on whether the cylinder actuating valves are of the memory type or use return springs, on the characteristics (such as fan-in) of the available logic elements, and finally, on whether we wish to lay more stress on circuit simplicity or on element economy. For the above reasons, Table 3.1 often indicates more than one preferred type of programmer for a given category of logic elements.

Note

1. For example, the program used here especially favours Johnson code programmers, for the following reason: In the circuit of Fig. 3.5 we would ordinarily need two additional AND gates for the output functions $A+ = y_1 y_2'$ and $A- = y_1' y_2$ (see single-state decoding functions listed in Table 2.6). However, in this particular program, it so happens that we can eliminate these two AND gates, and substitute the direct connections $A+ = y_1$ and $A- = y_1'$ shown in Fig. 3.5. This is so because the A+ signal, while only required for step No. 1, is permitted to remain for steps Nos. 1 to 3; and similarly the A− signal is permitted to remain for steps Nos. 4 to 6. Using the multi-state decoding rules listed in Section D of Chapter 2, we find that we happen to have $m = n$, so that no decoding gates are required. This situation, however, is not typical; i.e., in most programs we cannot expect to maintain each output signal for a number of steps exactly equal to the number of programmer flip-flops.

CHAPTER 4

PROGRAMMER OPERATING MODES

Three different aspects of programmer operating modes must be discussed: The first relates to resetting the programmer to its starting position after a power or other interruption. The second pertains to the way the programmer operation is started (START operating modes), while the third relates to what happens if the programmer is stopped before the program has been completely carried out (STOP operating modes). Some aspects of the problems involved are discussed in [9] and [11] to [16].

A. Resetting the programmer to starting position

The problem of resetting the programmer to its initial starting state may arise after the power (air supply or voltage supply) has been shut off. The problem is somewhat different when the flip-flops consist of valves, since these have a built-in mechanical memory and will remain indefinitely in their last position. When the air supply is renewed, a programmer using such valves will automatically continue the cycle at the point where it was interrupted. If, however, we wish the programmer to begin again with step No. 1, we must reset the programmer to its starting state.

The situation is otherwise with fluidic, moving-part fluidic, or electronic flip-flops. When power is renewed, the condition assumed by each flip-flop is a matter of chance. It is therefore always necessary after every power interruption to supply each flip-flop with a SET or RESET signal in order to place the programmer into its required starting state, so that the START signal will initiate a new cycle. This can easily be done by a single pressure line branching out to the various flip-flops. The required initial settings of the different flip-flops are summarized in Table 4.1 for different code programmers. In cases where the flip-flop inputs do not have a fan-in of 2 (or, for NOR gates, a fan-in of

3), this resetting pressure line must be connected to the single existing SET or RESET input of each flip-flop through an OR gate (e.g., a shuttle valve), so that the various inputs remain isolated from each other.

Table 4.1 Instructions for resetting programmer to starting position

Type programmer	First flip-flop	Intermediate flip-flops	Last flip-flop
1/*n* code	Reset	Reset	Set
2/*n* code	Set	Reset	Set
Accumulating code	Reset	–	–
Johnson code	Reset	Reset	Reset
(1,2)/*n* code	Set	Reset	Set

Note that, for the accumulating code programmer, it is sufficient to reset only the first flip-flop, as this will automatically reset all the remaining flip-flops (see Fig. 2.5). This will save additional connections, and also the need for the input OR gates where otherwise needed.

With some types of logic elements, it is possible to construct flip-flops having an internal bias so that, when the power supply is renewed, the flip-flop will automatically assume a defined state. With such elements, it would be possible to reset the programmer without any special reset arrangement, simply by interrupting and then renewing the power supply.

There are other types of flip-flops (e.g., those making use of feedback circuits as pictured in Figs. 3.9, 3.12 and 3.13), which will automatically assume a reset state when the power supply is renewed. For programmers using such flip-flops, it would only be necessary to supply the set signals specified in Table 4.1 after the power supply has been interrupted and renewed.

The proper start-up procedure for any programmer would be to first disconnect the power supply to all position sensors (i.e., limit valves, fluidic sensors, or limit switches), in order to prevent arbitrary motions of the various cylinders. Power is at first supplied only to the programmer itself, and the various flip-flops are set or reset according to the instructions of Table 4.1. Only then is the power supply reconnected to the position sensors, whereupon the START signal will be able to initiate the cycle.

A second aspect of the resetting procedure relates to the positions assumed by the various cylinder actuating valves. The problem arises if the programmer operation has been interrupted in the middle of the cycle. If we now wish to resume operation with step No. 1, it will not generally be sufficient to reset the programmer itself to its starting state, since some of the cylinder actuating valves having double pilot inputs may be in an incorrect position.

This problem is easily solved by connecting the programmer reset signal also to the required pilot lines of the actuating valves (through OR gates). Only pilot lines of valves having double pilot inputs need be considered. Valves with single pilot line and spring return will automatically assume their correct starting positions, since their pilot lines require sustained signals.

Using the programmer of Fig. 3.2 as an example, we would connect the programmer reset signal (coupled with the y_4 signal through an OR gate) to the A− pilot line, since this corresponds to the required starting position of cylinder A. Since cylinder B is actuated by a valve with spring return, no reset signal is necessary there. In addition, the reset signal would be connected (again through OR gates) to the following flip-flop inputs, according to Table 4.1: R_1, R_2, R_3, R_4, R_5, S_6.

In some systems (for example, certain machine tools or other manufacturing machines), it is necessary to have the various cylinders return to their starting position in a certain order, in order to avoid mechanical damage after a program interruption. This requirement, of course, complicates matters. One solution to this problem is to construct a secondary programmer which is actuated only by the reset signal of the primary programmer. The job of this secondary programmer would be to return the various cylinders to their respective starting positions in the required order. If a certain cylinder is already at its proper starting position, the corresponding step of the secondary program is automatically skipped. Thus, the proper order of cylinder motions is assured, regardless of the point at which the primary program was interrupted.

B. START operating modes

A number of START operating modes are discussed in [9] and [12]. While there are no universally accepted definitions of what is meant by the various START modes, the most commonly used modes seem to be the following:

1. *Automatic.* The program cycle is automatically and continuously repeated, until a stop signal is given.

2. *Semi-automatic.* Only one cycle is carried out. If the 'semi-automatic' signal is present at the end of the cycle, a new cycle is initiated.
3. *Single cycle.* A given 'single-cycle' signal will initiate one and only one cycle. For safety reasons, a new cycle can only be started if the next 'single-cycle' signal is actuated *after* the previous cycle has been completed.
4. *Semi-automatic with take-over.* For safety reasons, the start signal must be maintained manually (by pressing a button) until a certain step in the cycle is reached. From that point on, operation continues as for the semi-automatic mode.
5. *Single-cycle with take-over.* Same as semi-automatic with take-over, except that, after take-over, operation continues as for the single-cycle mode.
6. *Single-step operation.* Here, each program step is carried out separately, one at a time, by pressing a button. Useful for checking or adjusting operation of the system.

The implementation of the various START modes for different types of programmers will now be discussed.

1. Automatic

For all except accumulating code programmers, this is achieved simply by feeding the last input signal x_n back directly to the first programmer stage, *without* the use of the START input that was shown in Figs. 3.2, 3.3, 3.5 and 3.6. In this way, the cycle starts as soon as power is supplied to the programmer, and will continue to repeat itself indefinitely. If this feedback signal is interrupted by means of a normally-open valve, the programmer will stop at the end of the cycle. This is illustrated for a $1/n$ code programmer in Fig. 4.1.

For accumulating code programmers, the arrangement of Fig. 4.2 can be used. At the beginning of the cycle, $y_n = 0$ so that the $y_n' = 1$ signal sets flip-flop No. 1, initiating the cycle. During the last step of the cycle, $y_n = 1$ which extinguishes the SET signal y_n'. When the last step is completed, $y_n x_n = 1$, resetting flip-flop No. 1 and with it all other flip-flops. This, in turn, makes $y_n = 0$, automatically initiating the next cycle.

If the flip-flop should be of the type that gives $y = 0$ for $S = R = 1$ (i.e., RESET dominating), it would be possible to do without the NOT gate in Fig. 4.2, and instead feed a constant 1 signal into the SET input of the flip-flop. At the end of the cycle, we would have $y_n x_n = 1$, giving $S_1 = R_1 = 1$. As a result, this first

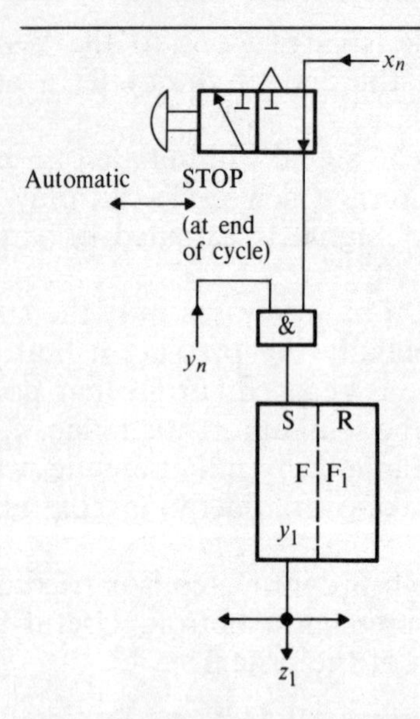

Fig. 4.1 Automatic START mode for $1/n$ code programmer

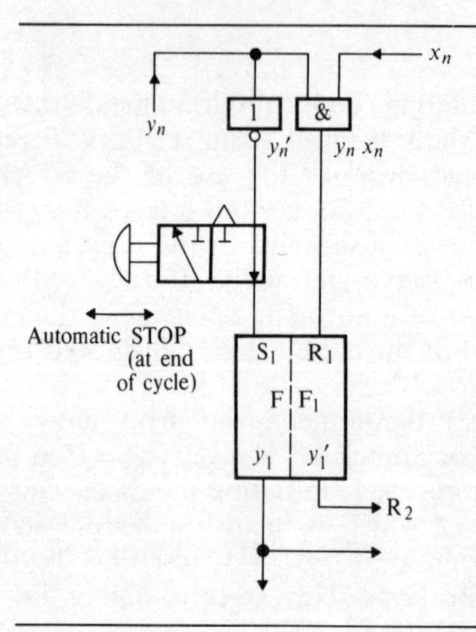

Fig. 4.2 Automatic START mode for accumulating code programmer

flip-flop would give $y_1' = 1$, providing the RESET signals for the remaining flip-flops. Once $y_n = 0$, we would get $S_1 = 1$ and $R_1 = 0$, initiating the next cycle. The flip-flops described in Figs. 3.9(a) and 3.14 are of the RESET dominating type.

2. *Semi-automatic*

This is almost identical to the Automatic START mode; indeed, the difference between these two modes is not much more than a matter of semantics: in the Automatic mode, the program is automatically repeated during the *absence* of a STOP signal, and is not repeated during the *presence* of the STOP signal. In the Semi-automatic mode, the program is automatically repeated during the *presence* of the START signal, and is not repeated during its *absence*. Thus, if we exchange the normally-open valve in Figs. 4.1 or 4.2 for a normally-closed valve with spring return, and change the name STOP to START, we have converted the circuits of Figs. 4.1 or 4.2 from Automatic to Semi-automatic operation. This is shown in Fig. 4.3 for a $1/n$ code programmer.

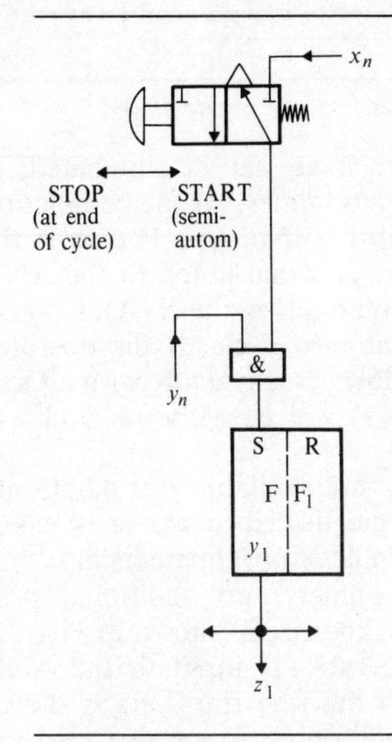

Fig. 4.3 Semi-automatic START mode for $1/n$ code programmer

3. *Single-cycle*

An elegant solution for this case is given in [11]. Two additional programmer stages are required at the end of the cycle, and their task is to prevent repetition of the cycle unless the START button is first released and then re-actuated. With a slight

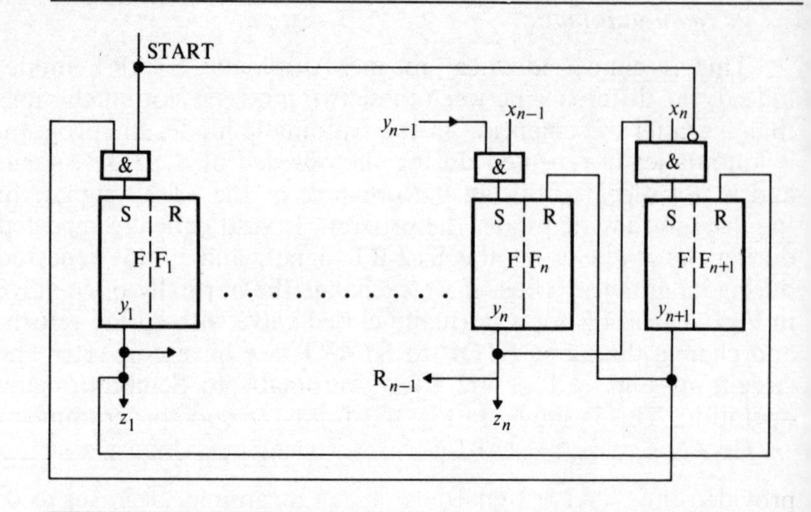

Fig. 4.4 Single-cycle START mode for $1/n$ code programmer

modification, one of these two extra stages can be eliminated, as shown in Fig. 4.4 for a $1/n$ code programmer. At the completion of the cycle, signal $y_n x_n$ will set flip-flop No. $(n+1)$ only if the START signal is 0. The resulting y_{n+1} signal is fed to the AND gate of the first flip-flop, so that renewal of the START signal will set this flip-flop and initiate the next cycle. If the operator attempts to tie the START button down permanently with a piece of tape or wire, flip-flop No. $(n+1)$ will never be set and will block the next cycle.

If AND gates with fan-in = 3 are not available, one additional AND gate is required. The principle described above is easily applied to $2/n$, Johnson, and $(1,2)/n$ code programmers also.

For accumulating code programmers, no additional programmer stages are needed. Instead, the circuit shown in Fig. 2.5 is slightly modified by adding a (START)′ input to the AND gate supplying the RESET signal of the first flip-flop, as shown in Fig. 4.5. The various flip-flops will only be reset provided the START signal is 0 at the end of the cycle when $y_n x_n = 1$. Restoring the START = 1 signal will initiate the next cycle.

For accumulating code programmers implemented with logic gates as illustrated in Fig. 3.10, the single-cycle START mode can be obtained by replacing the START signal with an OR gate giving the function $(\text{START} + y_n' + x_n')$. Until the end of the cycle, either y_n' or x_n' will be 1. At the end of the cycle, $y_n = x_n = 1$, so that the flip-flop feedback circuits will all collapse

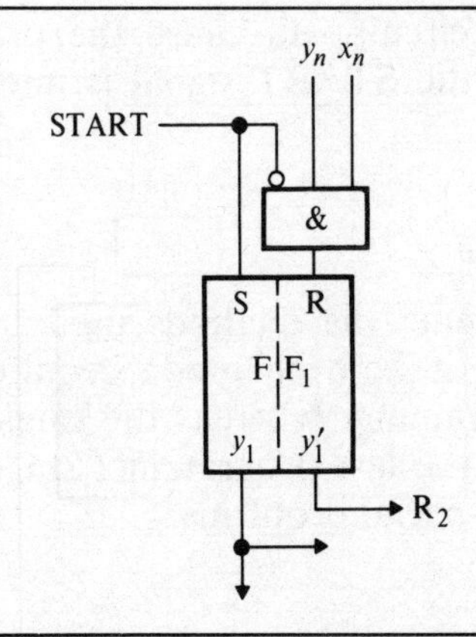

Fig. 4.5 Single-cycle START mode for accumulating code programmer

provided the START signal has, in the meantime, been set to 0. Renewing the START = 1 signal initiates the next cycle. With a constant START = 1 signal, all programmer stages remain indefinitely in the set condition. A different arrangement is suggested in [9] and [12].

4. Semi-automatic with take-over

This START mode can be obtained for the various types of programmers by adding the START signal as a third input to each one of the AND gates up to the programmer stage where take-over is to occur. (If AND gates with fan-in = 3 are not available, additional AND gates will be required.) For example, referring to Fig. 3.2, which shows a $1/n$ code programmer for the sample problem of Fig. 3.1, assume we require that the START button be held down until the completion of the first B+ stroke. We would therefore add the START signal as a third input to the AND gates of the second and third flip-flops. The $a+$ and $b+$ signals will now be unable to set their respective flip-flops unless the START button is pressed at the same time. Once the B+ stroke has been completed and flip-flop No. 3 set, the START button can be released, and the cycle will continue as for the semi-automatic mode. Interruption of the START signal before the take-over point will merely stop the program. When the START signal is renewed, the program will continue where it left off.

For accumulating code programmers implemented with logic gates as illustrated in Fig. 3.10, it is possible to use a different circuit shown in [9] and [12]. This circuit will cause the programmer to return to Step No. 1 if the START signal is interrupted before the take-over point.

5. *Single-cycle with take-over*

This mode is obtained by combining the methods used for the previous two modes. Since the take-over mode requires changes at the beginning of the programmer whereas the single-cycle mode requires modifications at the last programmer stage, combining these two modes poses no special problems.

6. *Single-step operation*

This mode is useful to permit adjustment of the machine or controlled system, or to diagnose faults or defects. Several possible methods are suggested in [11], but some of these are only suitable for specific types of logic elements. For example, it is suggested to cut off the regular air supply to the programmer stages, and then introduce short air-supply pulses through a special push-button valve each time an additional program step is desired. Obviously, this method will only work with Category A elements; i.e., valves or equivalent moving-part fluidic elements where the flip-flops possess mechanical memory and will remain in their last position when the air supply is cut off. Also, this method is not suitable if sustained output signals z_i should be required.

A more general solution suggested in [11] is to use a line connected as third input to the AND gate of each programmer stage. (Again, if AND gates with fan-in = 3 are not available, an additional AND gate per stage would be required.) As long as this line is supplied with air, the programmer will operate normally. If the air supply to this line is cut off, the programmer will stop. Introducing a short pulse into this line will permit that AND gate having the appropriate y_i and x_i inputs to set the next flip-flop and thus initiate the next step. Care must be taken to keep the pulses short enough to avoid initiating two or more successive program steps at a time (in cases where the actuated cylinder completes its stroke very quickly). This method will produce sustained outputs z_i only for other than $2/n$ or $(1,2)/n$ code programmers. With $2/n$ or $(1,2)/n$ codes, however, the outputs z_i will be pulses of the same duration as the pulses introduced into the AND gates. (If sustained output signals are required with such programmers, an additional flip-flop could here be used for each such output, or the flip-flop y_i outputs could be utilized, as indicated in Tables 2.4 and 2.8 respectively.)

A variation of the above method would be to cut off the regular air supply to the position sensing limit valves or fluidic sensors, and then introduce short air-supply pulses each time an additional program step is desired. This will enable the next input signal x_i to reach the relevant programmer stage and thus initiate the next step. This method avoids the need for additional AND gates. However, the above-mentioned considerations concerning the pulse length and the duration of the output z_i signals apply here also.

Other variations are, of course, possible, such as regulating the air supply to each position sensor (or the connection between sensor and AND gate) individually with separate push-button valves. This avoids the need to keep the pulses short. However, the operation is somewhat complicated in that a different push-button must be pressed for every single program step.

C. STOP–RESTART circuits

Before discussing the various STOP operating modes, some remarks about STOP–RESTART circuits and about the interaction between the RESTART and STOP signals are in order.

The inputs and outputs of a STOP–RESTART circuit are shown in the block diagram of Fig. 4.6. The STOP signal is applied in order to stop or interrupt the programmer cycle at any

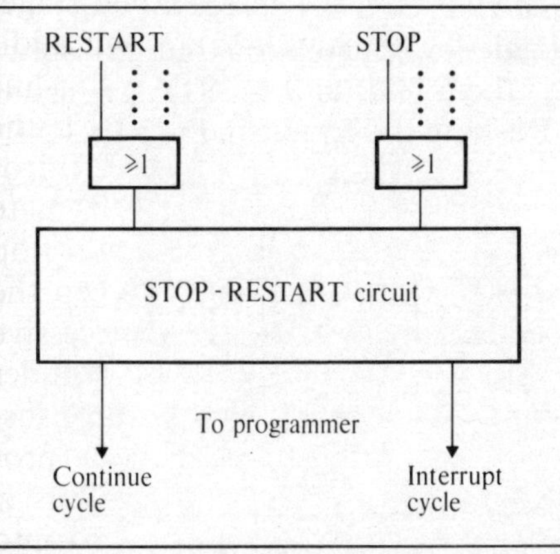

Fig. 4.6 Block diagram of STOP–RESTART circuit

stage before its completion. The RESTART signal is applied to continue the cycle after such interruption. It is important not to confuse this RESTART signal with the START signals dealt with up to now. These START signals were intended to make the programmer begin a new cycle after the completion of the previous one. If such a START signal is 0, the programmer will automatically stop at the end of the cycle, as was discussed in detail in connection with the various START operating modes. In other words, the presence or absence of a START signal will affect the programmer operation only at the completion of the previous cycle, whereas the RESTART signal must be able to take effect at any time during the cycle after the cycle has been interrupted by a STOP signal. We shall therefore deliberately use the term RESTART in order to differentiate this from the previously used START signal.

For convenience and safety reasons, it is often required that the above STOP and RESTART signals need only be short pulses; nevertheless, the programmer itself must receive sustained signals. It is the task of the STOP–RESTART circuit to convert the STOP and RESTART pulses received at its input into sustained signals. In order to differentiate between these input and output signals, we shall introduce two new terms: a RESTART pulse will produce a sustained 'Continue Cycle' signal, whereas a STOP pulse will eliminate the 'Continue Cycle' signal and may, if necessary, also produce a sustained 'Interrupt Cycle' signal. Not all STOP operating modes will require an explicit 'Interrupt Cycle' signal; i.e., it will often be sufficient to eliminate the 'Continue Cycle' signal in order to interrupt the cycle.

The simplest STOP–RESTART circuit consists of only one 3/2 valve with handle or toggle-lever actuation but no spring return (see Fig. 4.7(a)). Here, the STOP and RESTART signals consist of mechanical inputs. Pushing the handle in the RESTART

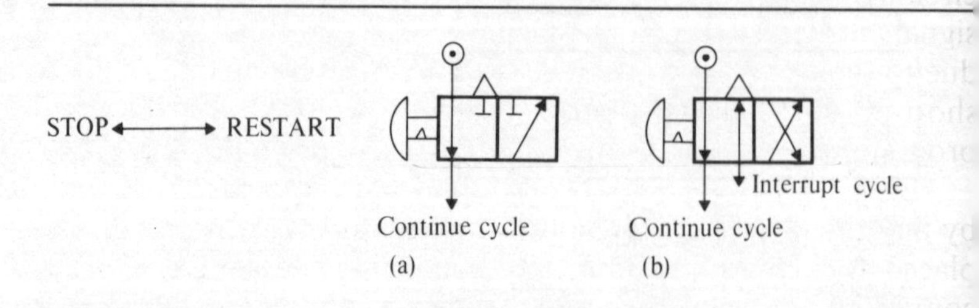

Fig. 4.7 Use of 3/2 or 4/2 valve without return spring as STOP–RESTART circuit

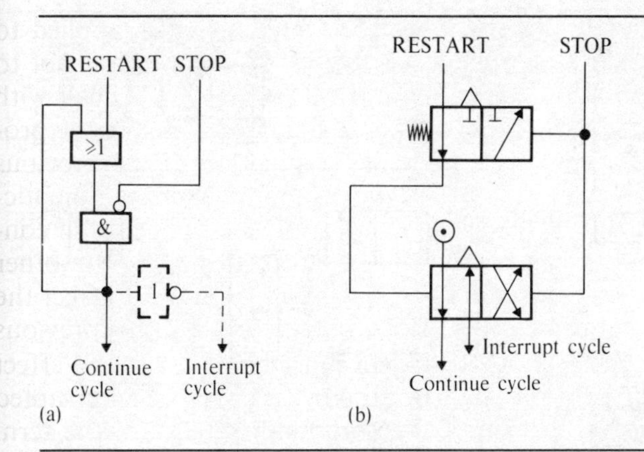

Fig. 4.8 STOP–RESTART circuit with STOP signal dominating (a) using gates in feedback-circuit flip-flop, (b) using valve as mechanical flip-flop

direction produces a 'Continue Cycle' signal; returning the handle in the STOP direction eliminates this signal. If an explicit 'Interrupt Cycle' signal is required, a 4/2 or 5/2 valve would have to be used, as shown in Fig. 4.7(b). The signals supplied by either of these valves could then be utilized in various ways, depending on the STOP mode required, as discussed later.

In many applications, however, it must also be possible to produce STOP and RESTART signals from several different locations (for instance, from the control panel but also from various locations on the machine itself), as indicated by the dotted inputs to the OR gates in Fig. 4.6. This requires the use of a logic circuit with memory, so that short STOP or RESTART pulses can control the circuit output. Furthermore, for safety reasons, the STOP signal must dominate if both STOP and RESTART signals would be actuated simultaneously (see [9] and [12]). This can be achieved by using one of the circuits of Fig. 4.8. For both circuits, a RESTART pulse will produce a 'Continue Cycle' signal, provided the STOP signal is 0 at the time. Furthermore, due to the use of a flip-flop, the RESTART signal need only be a short pulse, yet a sustained 'Continue Cycle' signal will reach the programmer, as long as the STOP signal is not actuated.

Both RESTART and STOP signals can, if desired, be supplied by any one of several 3/2 push-button valves with spring return, placed at different locations in the machine and with their outputs connected together through OR gates (though only one such RESTART and one STOP signal input has been drawn in Figs. 4.8 and 4.9).

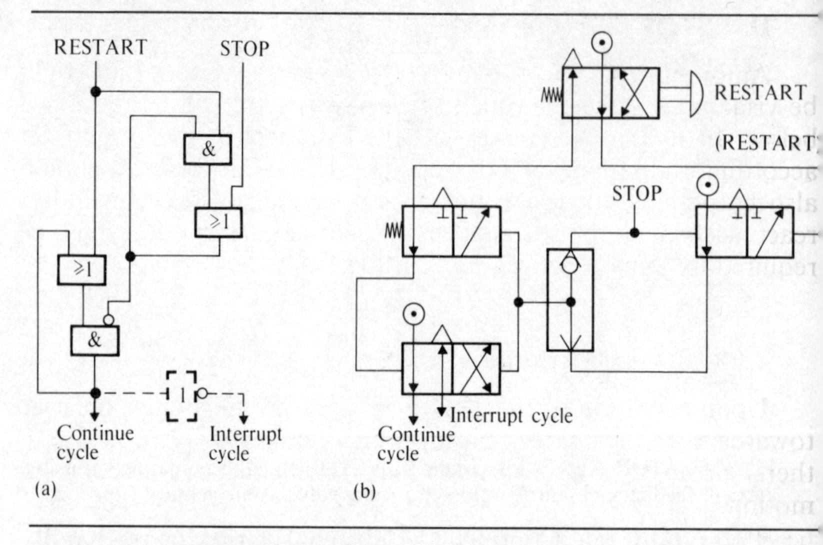

Fig. 4.9 STOP–RESTART circuit with momentary STOP signal dominating over substained RESTART signal (a) using gates in feedback-circuit flip-flops, (b) using valves as mechanical flip-flops

The above circuits of Fig. 4.8 suffer from one weakness: if the RESTART signal is maintained continuously (e.g., by taping or tying down the RESTART button), a continuous STOP signal will be required to counteract it. This may, in certain cases, be undesirable for safety reasons. In the two circuits of Fig. 4.9, a momentary STOP signal is sufficient to deactivate even a continuous RESTART signal. After the STOP signal has appeared it is first necessary to switch the RESTART signal to 0 before a renewed RESTART = 1 will be able to reactivate the 'Continue Cycle' signal. (Note that, in Fig. 4.9(b), an explicit (RESTART) signal is needed. This is obtained by using a 4/2 or 5/2 push-button valve, rather than a 3/2 valve. This 4/2 valve has therefore been included in this drawing, even though the push-button valves producing all the other RESTART and STOP signals have not been drawn in Figs. 4.8 and 4.9.)

The circuits shown in Figs. 4.7, 4.8 and 4.9 can be utilized in connection with the various STOP operating modes to be discussed in the following section. In cases where an explicit 'Interrupt Cycle' signal is required, the additional NOT gate shown in dotted lines in Figs. 4.8(a) and 4.9(a) is used. If no explicit 'Interrupt Cycle' signal is needed, this gate is dispensed with, and similarly a 3/2 valve can replace the 4/2 valve shown in Figs. 4.8(b) and 4.9(b).

D. STOP operating modes

Among the many different STOP operating modes that could be visualized, the eight which seem most useful will be discussed below. In the simplest cases, all the cylinders in the system act according to the same STOP operating mode. However, it should also be possible to design the system so that different cylinders react according to different STOP modes, since this may be required in many practical applications.

1. 'Go to Safety Position'

Upon actuation of the STOP signal, a given cylinder must go towards a defined safety position and remain there (unless it is there already), even if this means reversing the direction of motion.

This type of STOP mode is relatively expensive to implement, since each cylinder operating in this mode needs at least one additional valve. The methods described here will apply to any type programmer, regardless of the code used.

In the solution of Fig. 4.10, this STOP mode is obtained by actuating each cylinder that is required to go to a safety position by means of a 5/2 actuating valve with spring return (rather than with a valve having two pilot inputs). The single pilot input of this actuating valve gets its air from the appropriate z_i programmer output through a 3/2 valve with spring return. As long as a 'Continue Cycle' signal is present, this 3/2 valve will pass whatever z_i signal arrives from the programmer, and the cylinder will carry out the steps required by the program. As soon as the STOP button is pressed, the 'Continue Cycle' signal disappears, the pilot line of the 5/2 valve is vented through the 3/2 valve, and the cylinder automatically goes to the required safety position, if it is not already there. An explicit 'Interrupt Cycle' signal is not required for this solution.

Renewal of the 'Continue Cycle' signal will again connect the pilot line of each 5/2 valve to its respective z_i programmer signal, so that the cycle will continue where it left off. However, if one or more of the cylinder motions carried out (e.g., manually) during the STOP interval coincide with the succeeding required program steps, the programmer stages corresponding to these steps will automatically be set, and these cylinder motions will not be repeated a second time. If it is desired to 'freeze' the programmer during the STOP interval, it would be necessary to

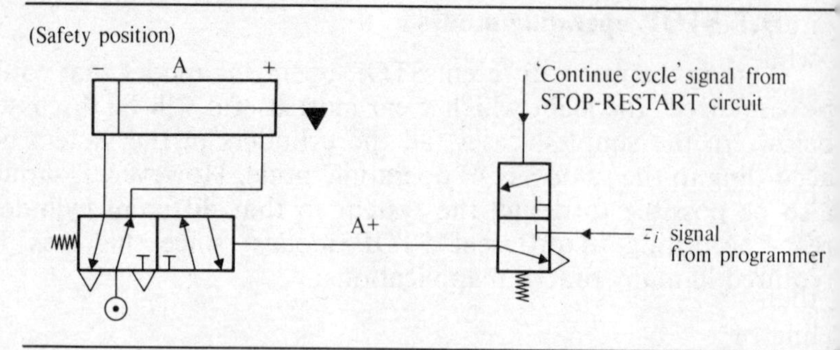

Fig. 4.10 Arrangement for sending cylinder to safety position upon actuation of STOP signal

interrupt the air supply to the position sensors before interrupting the 'Continue Cycle' signal. To complete the cycle, the 'Continue Cycle' signal would first be renewed, and only then would the air supply to the position sensors be reconnected. In this fashion, the programmer will continue exactly where it was interrupted, even if one or more steps have to be repeated.

If it is desired to start the cycle again from the very beginning after an interruption, it would be necessary to reset the programmer, as discussed in Section A of this chapter.

It should be noted that the 3/2 valve in Fig. 4.10 can be dispensed with in the case where the programmer makes use of low-pressure fluidic elements. This is because such elements require a pressure amplifier at each z_i output signal, and these pressure amplifiers can take the place of the 3/2 valves. It should only be necessary to interchange the input signals shown in Fig. 4.10;

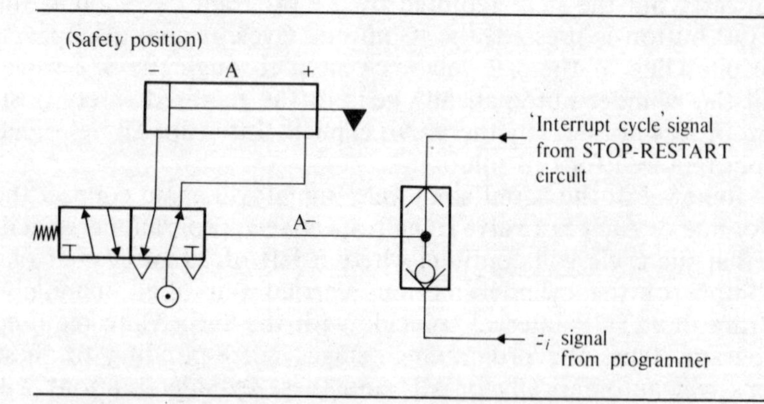

Fig. 4.11 Alternate arrangement for sending cylinder to safety position upon actuation of STOP signal

i.e., the z_i signal would act as input (pilot) signal to the amplifier, while the 'Continue Cycle' signal would take the place of the supply pressure to the amplifier.

Figure 4.11 shows an alternate arrangement for sending a cylinder to a defined safety position. Here, the connections between cylinder and actuating valve are reversed, and the pilot signal A− is obtained either from z_i or from an explicit 'Interrupt Cycle' signal connected through a shuttle valve (OR gate), rather than through a more expensive 3/2 valve (AND gate). The pilot line of the valve now represents A− rather than A+, and the programmer must therefore be connected to give $z_i = 0$ instead of $z_i = 1$ for A+ = 1, and vice versa.

Both Figs. 4.10 and 4.11 require a programmer design which provides sustained z_i signals. Figure 4.12 shows a second alternate arrangement, one which does not require sustained z_i signals. Here, an actuating valve with double pilot inputs is used. This avoids the need for sustained z_i signals, but requires both an additional shuttle valve (or other OR gate) and an additional 3/2 valve per cylinder. When the explicit 'Interrupt Cycle' signal is 0, the z_i and z_j signals from the programmer can reach the actuating valve.

2. 'No Change'

Upon actuation of the STOP signal, a given cylinder, if at rest, is to remain at its present position. If in motion, the cylinder is to maintain the motion until reaching the limit position and then remain there.

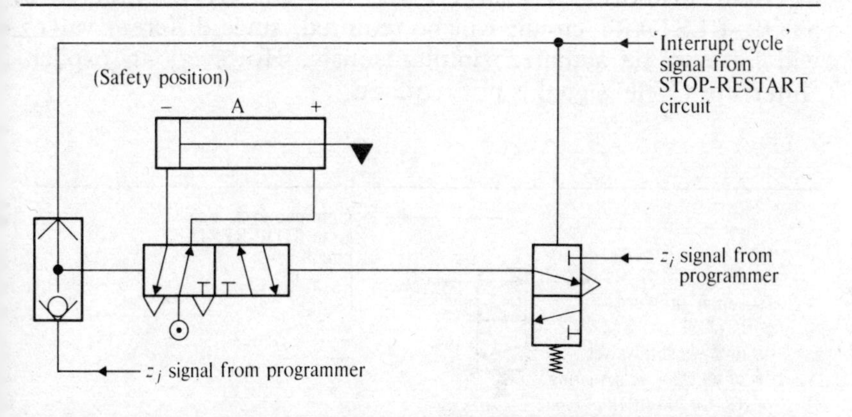

Fig. 4.12 Second alternate arrangement for sending cylinder to safety position upon actuation of STOP signal

This STOP mode is achieved very simply, for all except $2/n$ and $(1,2)/n$ code programmers, by interrupting the air supply to the limit or position-sensing valves, as shown in Fig. 4.13. Upon interruption of the 'Continue Cycle' signal from the STOP–RESTART circuit, the supply-pressure line feeding the various position-sensing valves is vented to atmosphere. As a result, the x_i feedback signals to the programmer are interrupted, so that the cycle stops at its present step. Every cylinder remains where it happens to be or, if in motion, completes its present motion. Upon renewal of the 'Continue Cycle' signal, the x_i feedback signals again reach the programmer, so that the cycle will continue where interrupted.

For $2/n$ and $(1,2)/n$ code programmers, the above method will only be suitable if no sustained z_i signals are required, since all z_i signals will disappear the moment the air supply to the position sensors is cut off. If sustained z_i signals are required with such programmers, one of the solutions presented under the START mode No. 6 'Single-step operation' should be used. Altogether, this START mode is really identical to the 'No Change' STOP mode, and all comments made there apply here also.

The 3/2 valve in Fig. 4.13 can, of course, also be operated directly by means of a push button (using a valve with no spring return), rather than by means of a pilot pressure. In that case, the STOP–RESTART circuit assumed to supply the pilot pressure can be dispensed with. This, however, presupposes that all cylinders in the system are subject to the same STOP mode. If different cylinders are subject to different STOP modes, a STOP–RESTART circuit will be required, since different valves will have to be actuated simultaneously. However, an explicit 'Interrupt Cycle' signal is not required.

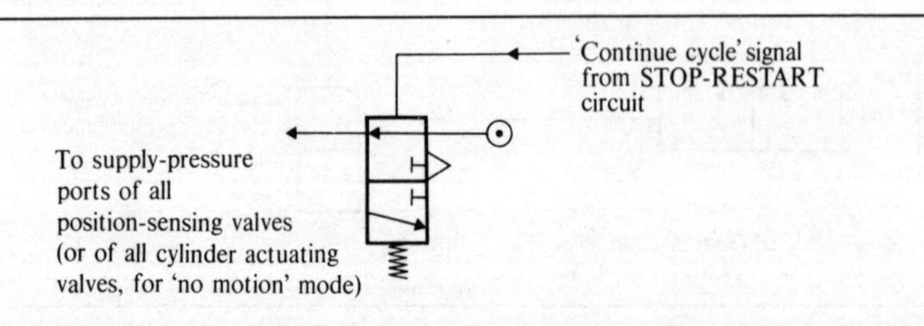

Fig. 4.13 Arrangement for 'No Change' (or 'No Motion') STOP mode

3. 'No Motion'

Upon actuation of the STOP signal, pressure on both sides of the piston is to be released, so that the piston motion will cease, unless some outside force acts on the piston.

This STOP mode is obtained by interrupting the air supply to the various cylinder actuating valves. The 3/2 valve required for this purpose is identical to that shown in Fig. 4.13, except that now the purpose of this valve is to cut off the air supply to the cylinder actuating valves, rather than to the position-sensing valves. As soon as this is done, the pressures on both sides of all cylinders are automatically vented, so that the cylinders will stop their motion, except for inertia effects. The cylinder pistons can now be moved manually, if so desired.

Upon renewal of the air supply to the cylinder actuating valves, the cycle will continue where interrupted. Since this method does not interfere in any way with the programmer itself, it can be used with any type programmer.

An explicit 'Interrupt Cycle' signal is not required, and, as explained in connection with the 'No Change' STOP mode, it may be possible to dispense with the STOP–RESTART circuit altogether.

The method of Fig. 4.13 described above may not work properly in cases where the cylinder velocity is limited by means of a restriction in the exhaust line ('Meter-out' circuit). This is

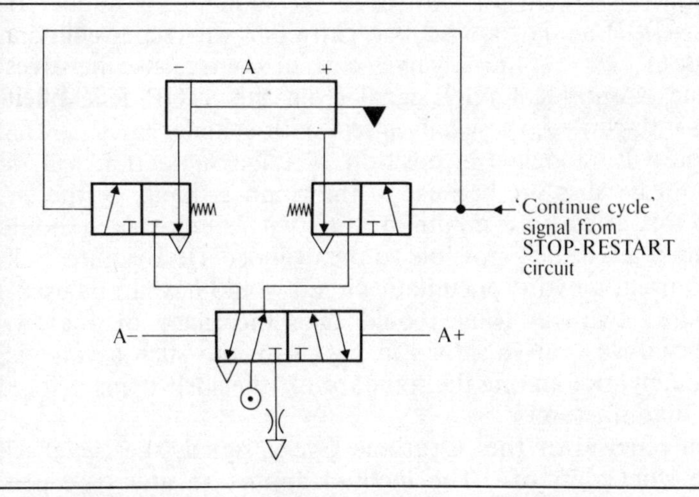

Fig. 4.14 Alternate circuit for 'No Motion' STOP mode

because, at the moment the air supply to the cylinder is cut off, there may still be considerable accumulated pressure in the cylinder exhaust side. This pressure is unable to escape quickly enough because of the restriction, and is liable to drive the piston back into its previous position. In cases where the cylinder drives a load with considerable inertia, the above action of the restriction may actually be useful in helping to decelerate the piston. However, where the action of the restriction is considered undesirable, we have two options:

One option is to place the restriction in the air-supply line (i.e. 'Meter-in' circuit), between the 3/2 valve and the cylinder actuating valve. This option also provides the useful side-benefit that the cylinder speed will be restricted the moment the air supply is renewed (which is not the case with the 'Meter-out' circuit after the pressure on the exhaust side has been released).

If the well-known drawbacks of 'Meter-in' velocity control are found too objectionable, we can use the second option, namely a circuit using two additional 3/2 valves, as shown in Fig. 4.14. Here, both cylinder sides are immediately vented the moment the 'Continue Cycle' signal is interrupted, and the exhaust restriction will have no effect.

4. 'Lock Piston'

Upon actuation of the STOP signal, the piston is to be locked in position.

This STOP mode requires two extra 2/2 valves per cylinder, as shown in Fig. 4.15. (3/2 valves can, of course, also be used.) When the 'Continue Cycle' signal from the STOP–RESTART circuit is interrupted, these valves seal both cylinder lines, so that the piston will be locked in position. The locking action will, of course, not be absolute because of the compressibility of the air. To keep this effect to a minimum, the two sealing valves should be mounted as close as possible to the cylinder. (If absolute locking is required, a hydro-pneumatic circuit would have to be used.) One double two-way valve could take the place of the two separate two-way valves shown in Fig. 4.15, but such a valve is, unfortunately, not among the standard off-the-shelf items offered by most manufacturers.

Upon renewal of the 'Continue Cycle' signal, the cycle will continue where left off. The method applies to any type programmer. An explicit 'Interrupt Cycle' signal is not needed.

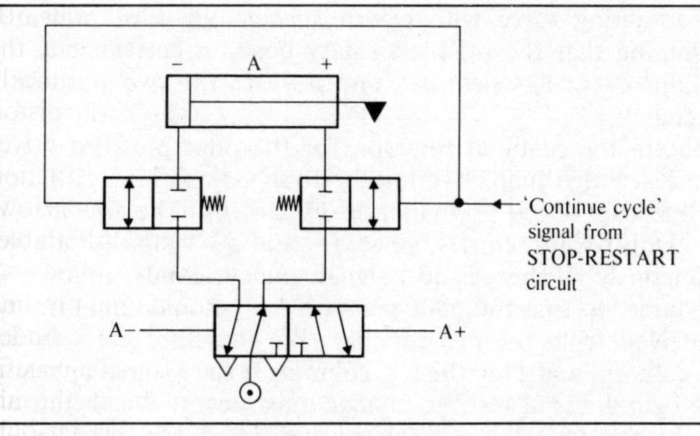

Fig. 4.15 Circuit for 'Lock Piston' STOP mode

The above four STOP modes discussed so far can be applied to any specific cylinder, or even to all the cylinders in the system. However, we often require STOP modes which will only apply to those cylinders that happen to be in motion; i.e., any cylinder at rest when the STOP signal is actuated should remain at rest, and only those cylinders that are in motion should be subject to the STOP mode. We shall denote such STOP modes as 'No Change/– –' modes, where the 'No Change' applies to those cylinders at rest, and the condition substituted for the two dashes to those cylinders that are in motion.

The following four 'No Change/– –' STOP modes seem to have practical application, and will be discussed in detail:

'No Change/Safety Position'
'No Change/Reverse'
'No Change/No Motion'
'No Change/Lock Piston'

Of course, the 'No Change' STOP mode previously discussed also belongs to the 'No Change/– –' category, since it could have been called 'No Change/Maintain Motion'.

5. 'No Change/Safety Position'

Upon actuation of the STOP signal, the cylinder, if at rest, remains at its present position. If in motion, the cylinder must go to a defined safety position, even if this means reversing the direction of motion.

For the implementation of this STOP mode, we use a 5/2

cylinder actuating valve with return spring and pilot pressure A+, assuming that the required safety position corresponds to A+ = 0 and $a-$ = 1, where $a-$ and $a+$ are the two position-sensor signals.

To obtain the required function for the pilot pressure A+, we plot a Karnaugh map covering the four variables $a-$, $a+$, z_i and I ('Interrupt Cycle' signal) (see Fig. 4.16). The $(a-)(a+)$ row gets 'Don't Care' entries, since $a-$ and $a+$ cannot both be 1 simultaneously. If there is no I signal, the cycle must follow its regular course, so that the pilot pressure A+ should equal the z_i signal received from the programmer. We therefore enter 0s in the $\mathrm{I}'z_i'$ column, and 1s in the $\mathrm{I}'z_i$ column. If the I signal appears while the cylinder is at rest, no change must occur. We therefore enter 0s in the two $\mathrm{I}\cdot a-$ squares, and 1s in the two $\mathrm{I}\cdot a+$ squares. Finally, if the I signal appears while the cylinder is in motion, the cylinder must move to the safety position, which is assumed to correspond to A+ = 0. We therefore enter 0s in the two $\mathrm{I}(a-)'(a+)'$ squares. The resulting simplified function is

$$\mathrm{A}+ = \mathrm{I}'z_i + \mathrm{I}(a+) = (\text{Continue Cycle})z_i + (\text{Interrupt Cycle})(a+)$$

The resulting circuit is shown in Fig. 4.16, and can be implemented using three logic gates (or corresponding valves). A STOP–RESTART circuit providing an explicit 'Interrupt Cycle' signal in addition to the 'Continue Cycle' signal is required. The solution is suitable for any type of programmer, but requires sustained z_i signals.

6. *'No Change/Reverse'*

Upon actuation of the STOP signal, the cylinder, if at rest, remains at its present position. If in motion, the cylinder is to reverse direction and return to its previous position.

This STOP mode, which is very useful, is more difficult to obtain than the previous one. A circuit for implementing this mode is given in [14], but is not presented here since it requires one additional flip-flop and eight additional gates per cylinder, and is thus extremely costly. A much simpler solution is presented in [13] and [15], but is only suitable for accumulating code programmers.

A simple solution suitable for any type programmer and requiring only four additional gates is obtained by using the same technique used for the previous STOP mode. The Karnaugh-map entries are made using similar reasoning (see Fig. 4.17). The map

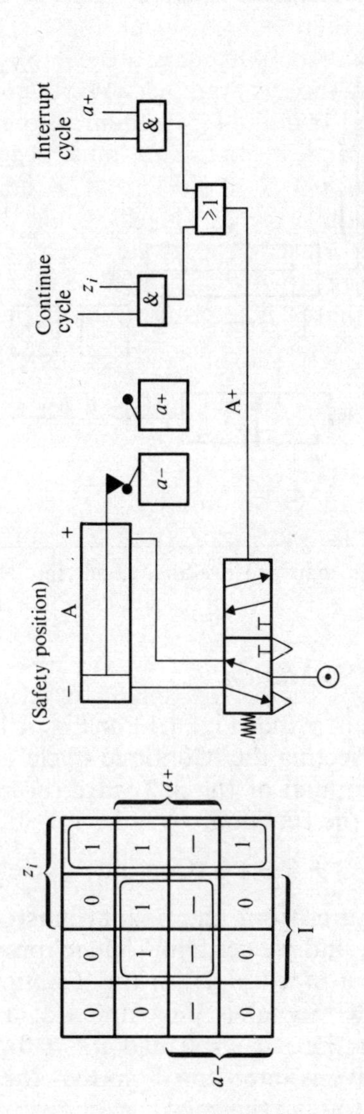

Fig. 4.16 Karnaugh map and resulting circuit for 'No change/Safety Position' STOP mode

is identical to that in Fig. 4.16, except for the 1-entry in the $(a-)'(a+)'\mathrm{I}z_i'$ square. The equation for A+ is

$$\begin{aligned} \mathrm{A}+ &= \mathrm{I}'z_i + \mathrm{I}(a+) + \mathrm{I}z_i'(a-)' \\ &= (\text{Continue Cycle})\,(z_i) + (\text{Interrupt Cycle})\,(a+) \\ &\quad + (\text{Interrupt Cycle})\,(z_i)'(a-)' \end{aligned}$$

The resulting circuit is shown in Fig. 4.17.

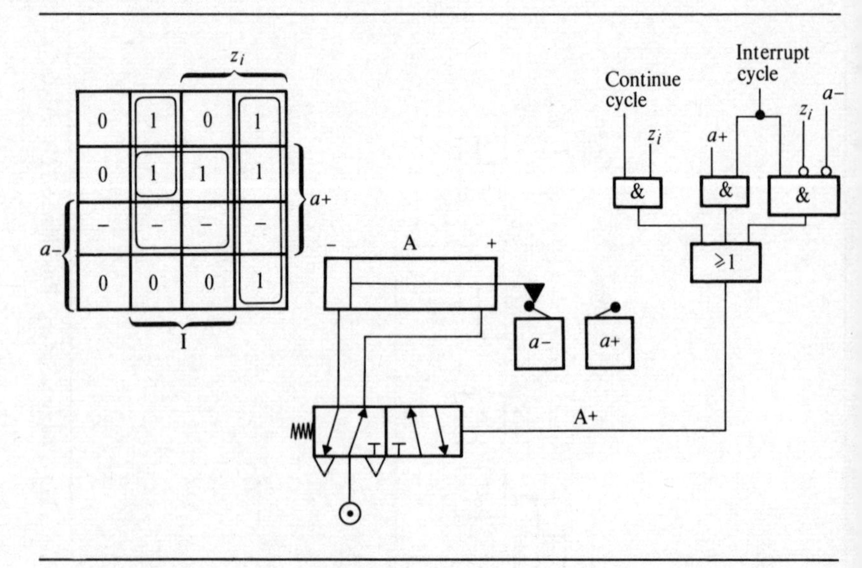

Fig. 4.17 Karnaugh map and resulting circuit for 'No Change/Reverse' STOP mode

7. 'No Change/No Motion'

Here, the circuit of either Fig. 4.13 or Fig. 4.14 is used. However, instead of connecting the 'Continue Cycle' signal directly to the pilot input (or inputs) of the 3/2 valve (or valves), the pilot input is supplied by the function

$$\text{Pilot input} = (a-) + (a+) + (\text{Continue Cycle})$$

Thus, the air pressure from the 5/2 cylinder actuating valve reaches the cylinder, and we get 'No Change' provided the piston is at rest (i.e., $a- = 1$ or $a+ = 1$), or the 'Continue Cycle' signal is given. If this latter signal is interrupted at a time when the piston is at motion (i.e., $a- = 0$ and $a+ = 0$), then the pilot signal to the 3/2 valve is interrupted, so that the air pressure on both sides of the piston is exhausted.

8. 'No Change/Lock Piston'

Here, the circuit of Fig. 4.15 is used, with the pilot input line being supplied by the function

Pilot input = $(a-) + (a+)$ + (Continue Cycle)

just as with the previous STOP mode.

CHAPTER 5

PROGRAMMERS FOR MULTI-PATH PROGRAMS

In this section, programmers for multi-path programs will be discussed, as opposed to the single-path programs dealt with so far. The various basic possibilities for multi-path programs are summarized in Fig. 5.1, where dots represent the states between successive program steps.

Figure 5.1(a) shows a single-path program, for comparison. Figure 5.1(b) represents a multi-path program with two simultaneous parallel paths. (More than two parallel paths are, of course, also possible.) The program runs through both parallel paths at the same time. Only when *both* parallel program branches have been completed does the programmer continue with the steps of the succeeding common path. This is achieved by connecting the proper input signals through an AND gate. Parallel-path programs of this type are, for example, required in multi-station machine tools where several different operations are to be carried out simultaneously. The workpiece must only be removed when *all* operations have been completed.

Figure 5.1(c) shows two alternative parallel paths. The programmer will select either one of these two paths, depending on whether a certain input signal p has been set $p = 1$ or $p = 0$. The final feedback input x signals of the two parallel branches are, in this case, connected through an OR gate, so that the program will continue when either one of the parallel program branches has been completed. A program of this type is suitable for multipurpose machines. The machine operator can select the branch to be actuated by setting p manually, or p could be set automatically according to some other outside condition.

Figure 5.1(d) shows a program where a number of program steps can be skipped, if desired. If we set $p = 1$, all steps are carried out successively. If, however, $p = 0$, the steps A_1 to A_j are skipped. It is obvious that this program really represents a special case of the parallel-path program of Fig. 5.1(c), with the

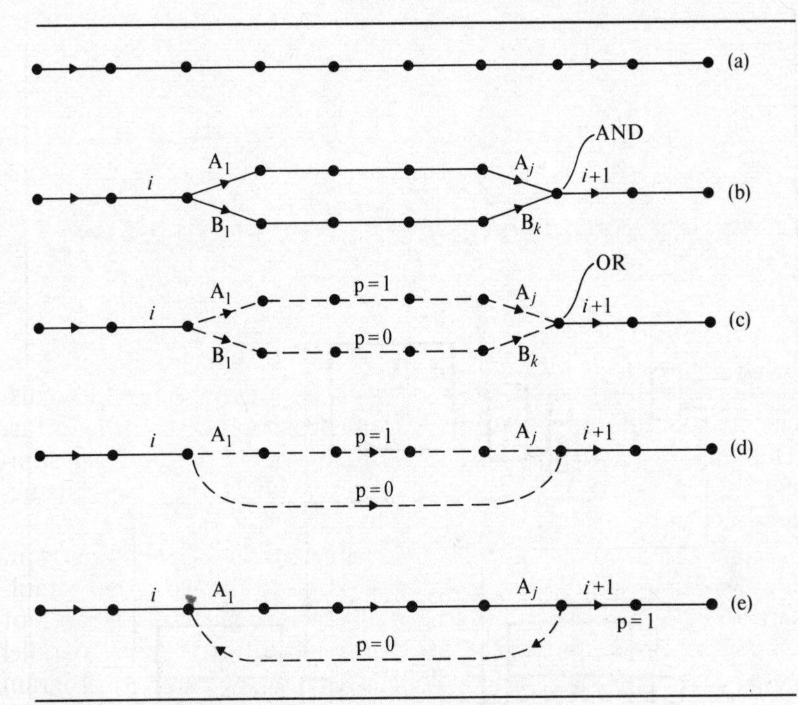

Fig. 5.1 Examples of single- and multi-path programs

lower p = 0 path having shrunken to a single step.

Finally, Fig. 5.1(e) describes a program where a number of steps can be repeated, if desired. If p = 1, the steps are not repeated. If p = 0, the steps from A_1 to A_j are repeated over and over again until p is reset to p = 1. This, too, represents a special case of the parallel-path program of Fig. 5.1(c). Both Figs. 5.1(d) and (e) represent programs suitable for multi-purpose machines, with p being set either manually or automatically.

We shall now discuss the implementation of these various programs using the different types of programmer codes.

A. Program with simultaneous parallel paths – Fig. 5.1(b)

Figure 5.2 shows the implementation using a $1/n$ code programmer. Completion of step i is signalled by $y_i x_i = 1$. This sets both flip-flops A_1 and B_1, actuating both parallel paths A and B. Completion of path A is indicated by $y_{A_j} x_{A_j} = 1$ and that of path B by $y_{B_k} x_{B_k} = 1$. (Since the two parallel paths are not necessarily of equal length, we need not have $j = k$.) These two signals are

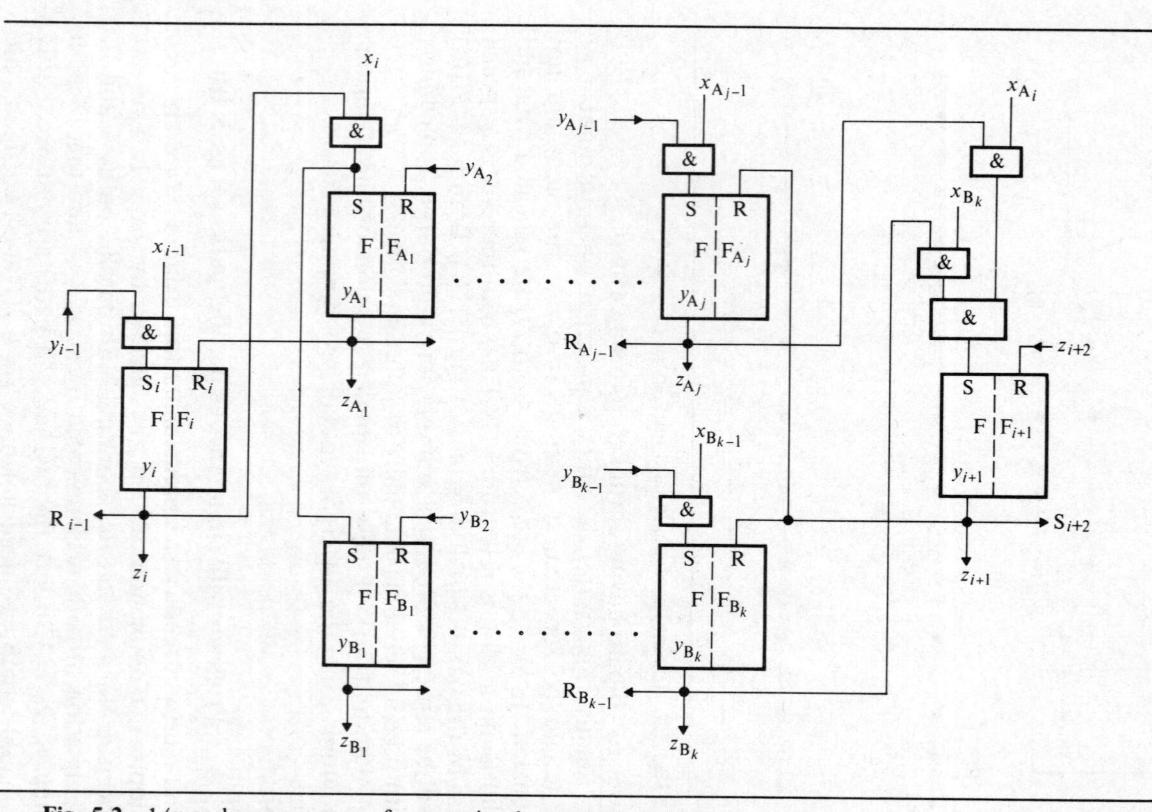

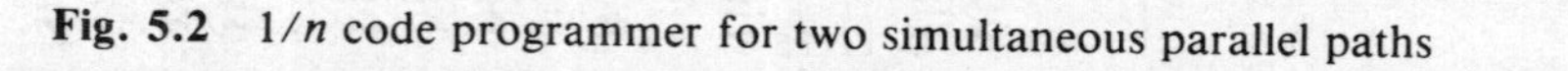

Fig. 5.2 $1/n$ code programmer for two simultaneous parallel paths

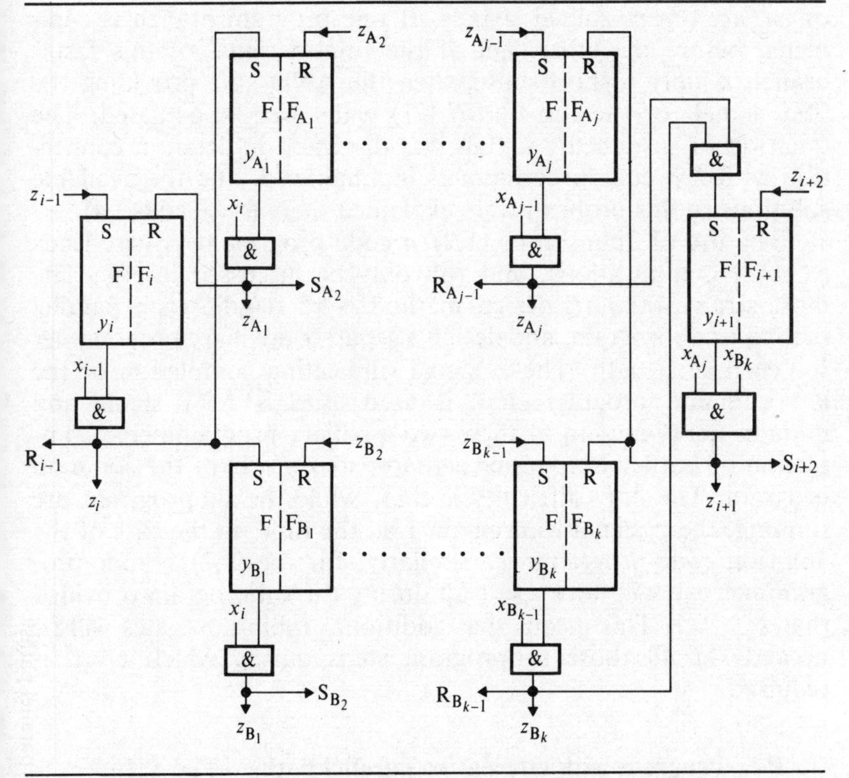

Fig. 5.3 2/*n* code programmer for two simultaneous parallel paths

connected through an AND gate, supplying the SET signal to flip-flop $(i+1)$ only after both paths are completed.[1]

If an accumulating code programmer is to be used, the circuit will be similar to that of Fig. 5.2, except for the RESET connections, which must be made as was illustrated in Fig. 2.5.

Figure 5.3 shows the implementation using a 2/*n* code programmer. Completion of step $(i-1)$ produces signal z_i, which sets both flip-flops A_1 and B_1. Completion of step i (signal x_i) produces output signals z_{A_1} and z_{B_1}, so that both parallel program branches are actuated simultaneously. When both of these branches are at their final steps, we get both output signals z_{A_j} and z_{B_k}, and these, connected through an AND gate, set the flip-flop $(i+1)$. Completion of both parallel branches produces the x_{A_j} and x_{B_k} signals, which, through an AND gate, provide the z_{i+1} output signal and thus initiate the next single-path program step.

The circuit of Fig. 5.3 may not operate properly if either z_{A_j}

or z_{B_k} are not sustained signals. If one program branch is completed before the other, and if the final z signal of this faster branch is only a short pulse, then the AND gate providing the SET signal for the flip-flop $(i+1)$ will never be actuated. The question of sustained z signals was discussed in detail in connection with $2/n$ code programmers in Chapter 2, and the available solutions to this problem were explained there (see Table 2.4).

The use of Johnson or $(1,2)/n$ code programmers here leads to some complications, and will only be discussed briefly. The most straightforward design method is to handle each parallel path as a subprogram, and design a separate auxiliary programmer for each such path. The x_i signal (indicating completion of the last common program step) is used as a START signal and initiates the operation of these two auxiliary programmers. Completion of both subprograms actuates step $(i+1)$ of the common program. The only difficulty is that, while the subprograms are running, the z_i signal will remain 1 all the time, in the case of the Johnson code programmer. Similarly, for the $(1,2)/n$ code programmer, we will have $z_i = 1$ all during the subprogram provided that $x_{i-1} = 1$. This means that additional Inhibition gates will be needed for all those subprogram steps during which $z_i = 0$ is required.

B. Program with alternative parallel paths – Fig. 5.1(c)

Figure 5.4 shows implementation using a $1/n$ code programmer. Completion of step i is signalled by $y_i x_i = 1$. If, at that moment, $p = 1$, flip-flop A_1 will be set and the program will continue according to the upper path. If, however, $p = 0$, flip-flop B_1 is set and the lower-path program is carried out. The $y_{A_j} x_{A_j}$ and $y_{B_k} x_{B_k}$ terms are connected through an OR gate, so that the completion of either path sets flip-flop $(i+1)$. Note that four additional gates are required (apart from the regular AND gates attached to each flip-flop), namely two OR gates, one AND gate, and one NOT gate for providing p'. (Two further additional AND gates may be needed if AND gates with fan-in = 3 are unavailable.)

If an accumulating code programmer is to be used, the circuit will be similar to that of Fig. 2.4, except for the RESET connections, which must be made as was indicated in Fig. 2.5. In addition, the flip-flop $(i+1)$ needs a RESET signal $R_{i+1} = y'_{A_j} y'_{B_k}$ to avoid its being reset before the completion of the program.

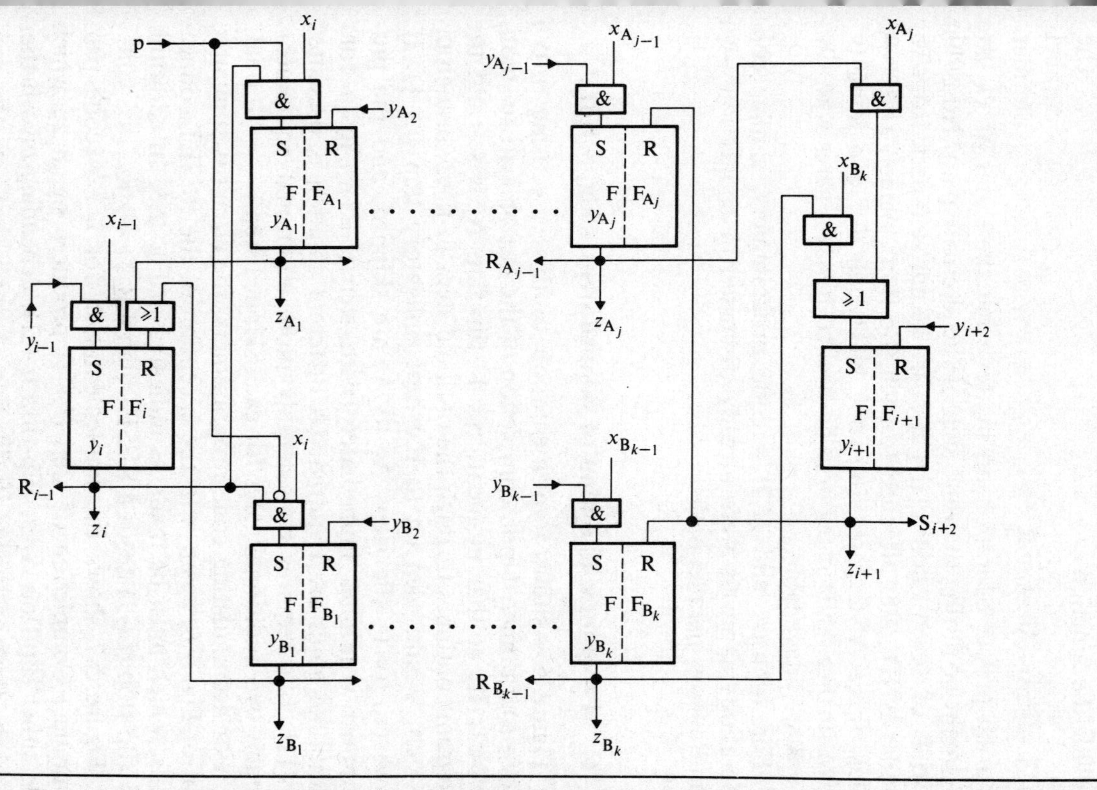

Fig. 5.4 $1/n$ code programmer for two alternative parallel paths (Path A for p = 1; Path B for p = 0)

Figure 5.5 shows implementation using a $2/n$ code programmer. Completion of step $(i-1)$ produces signal z_i, which sets either flip-flop A_1 or flip-flop B_1, depending on whether p equals 1 or 0 respectively. This determines which alternate path is to be followed. During the last step of the alternate path, either z_{A_j} or z_{B_k} becomes 1, setting flip-flop $(i+1)$. Completion of this last step produces $z_{i+1}=1$, so that the next single-path program step is initiated. Two AND gates and one OR gate are used to provide the function

$$z_{i+1} = y_{i+1}(\mathrm{p}x_{A_j} + \mathrm{p}'x_{B_k})$$

assuring proper operation no matter what the states of x_{A_j} and x_{B_k} might be during the final step of the alternate path. Note that a total of seven additional gates are required (apart from the regular AND gates attached to each flip-flop), namely three OR gates, three AND gates, and one NOT gate for providing p′. (Two further AND gates may be needed if AND gates with fan-in = 3 are unavailable.)

For Johnson and $(1,2)/n$ code programmers, similar considerations apply as were already discussed in connection with simultaneous parallel paths.

C. Program with option of skipped steps – Fig. 5.1(d)

Figure 5.6 shows implementation of this case using a $1/n$ code programmer. Upon completion of the last obligatory step i, $y_i x_i = 1$. If, at that moment, p = 1, flip-flop A_1 is set and the program continues through the optional steps of the A branch up to step A_j and then on to the next obligatory step $(i+1)$. If, however, p = 0, the steps A_1 to A_j are skipped, and the programmer continues immediately with step $(i+1)$. Again four additional gates are required. (A different solution is presented in [11], but requires considerably more additional equipment, namely two additional gates for each skipped step.)

For accumulating code programmers, the circuit is similar to that of Fig. 5.6, again with the exception of the RESET connections, which must be made as indicated in Fig. 2.5. In addition, the flip-flop $(i+1)$ needs a RESET signal $R_{i+1} = y_i' y_{A_j}'$.

Figure 5.7 shows the implementation for a $2/n$ code programmer. Completion of step $(i-1)$ produces signal z_i, which sets either flip-flop A_1 or flip-flop $(i+1)$, depending on whether p = 1 or 0 respectively. If flip-flop A_1 has been set, the pro-

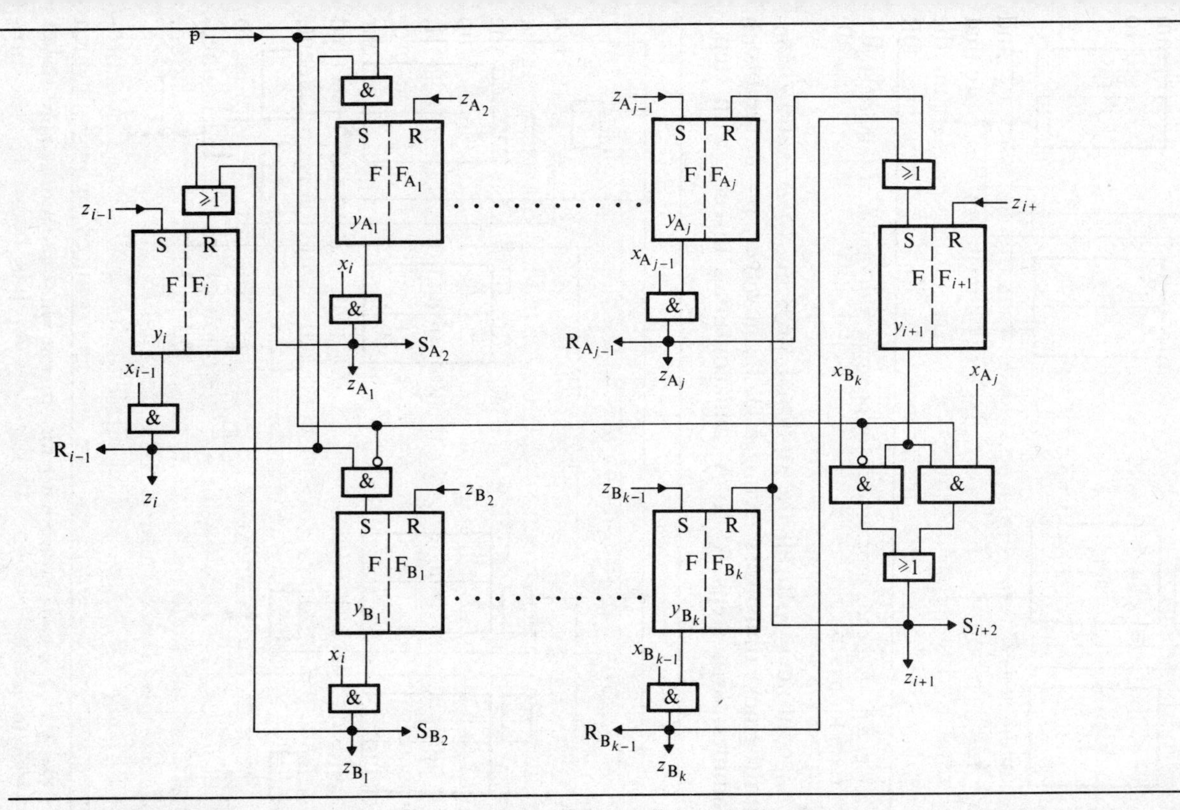

Fig. 5.5 2/*n* code programmer for two alternative parallel paths (Path A for p = 1; path B for p = 0)

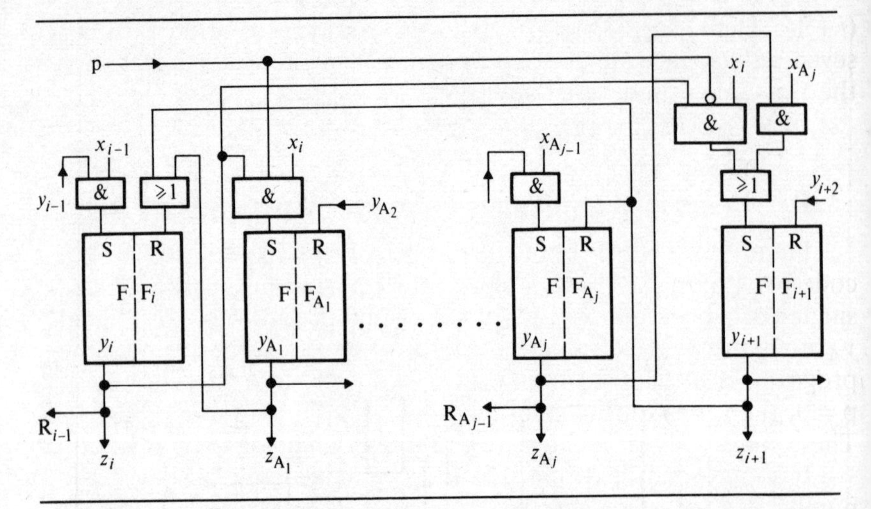

Fig. 5.6 $1/n$ code programmer for program with option of skipped steps (skip for p = 0)

gram continues with all the optional steps A_1 to A_j, and then with the next obligatory step $(i+1)$. If, however, $p = 0$, the programmer skips steps A_1 to A_j and continues directly with step

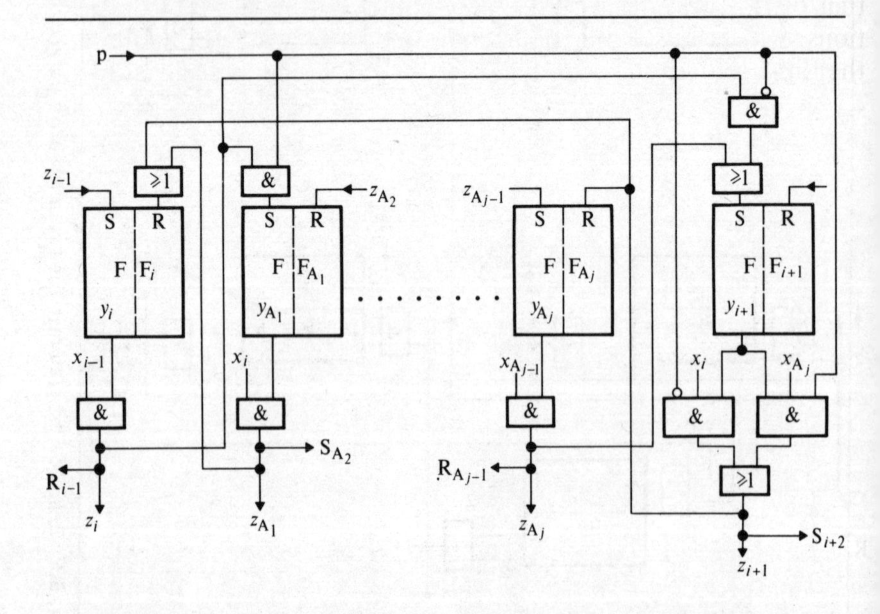

Fig. 5.7 $2/n$ code programmer for program with option of skipped steps (skip for p = 0)

$(i+1)$. Here, too, the $2/n$ code programmer solution requires seven additional gates, as compared to four additional gates for the $1/n$ code programmer solution.

D. Programmer with option of repeated steps – Fig. 5.1(e)

Figure 5.8 shows implementation of this case using a $1/n$ code programmer. Up to step A_j, the programmer operates as a single-path programmer. If, upon completion of step A_j (signal $y_{A_j}x_{A_j}$), the variable $p = 1$, then flip-flop $(i+1)$ is set, and the program continues without repeating any steps. If, however, $p = 0$, then flip-flop A_1 is set, and the steps A_1 to A_j are repeated. These steps will be repeated over and over again as long as $p = 0$.

It should be noted that, for the circuit to work properly, the number of repeated steps must be at least three (i.e., $j > 2$). If $j = 2$, the flip-flops A_1 and A_2 would each get SET and RESET signals simultaneously. In cases where the cycle to be repeated contains only two steps, it would be necessary to insert an additional 'dummy' stage into the programmer to make $j = 3$.

For accumulating code programmers, the circuit is similar to that of Fig. 5.8, again with the exception of the RESET connections, which must be made as indicated in Fig. 2.5. In addition, the flip-flop A_1 will require SET and RESET signals as given by

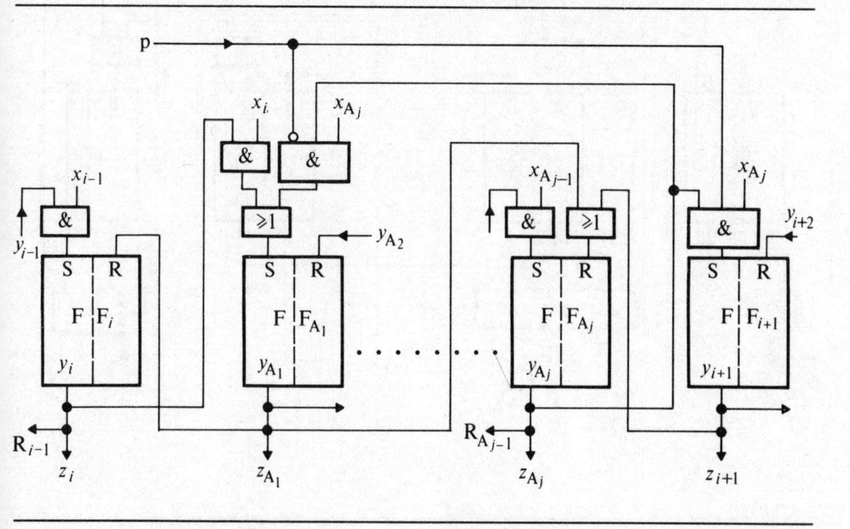

Fig. 5.8 $1/n$ code programmer for program with option of repeated steps (repeat for $p = 0$)

$$S_{A_1} = y_i R'_{A_1}(x_i + x_{A_j} p')$$
$$R_{A_1} = y'_i + y_{A_j} x_{A_j} p'$$

In this way, upon completion of step A_j, flip-flop A_1 is reset (provided $p = 0$), and with it all flip-flops A_1 to A_j. During this time, the S_{A_1} signal is inhibited. When flip-flop A_j has been reset, $y_{A_j} = 0$ so that $R_{A_1} = 0$ and $S_{A_1} = 1$, and the repeated cycle begins anew.

Figure 5.9 shows the implementation using a $2/n$ code programmer. Completion of step $(i-1)$ produces signal z_i, which sets flip-flop A_1. Completion of step i produces z_{A_1} through the AND gate having output $y_{A_1} x_i y'_{A_j}$. The program then continues up to completion of step A_{j-1} and appearance of signal z_{A_j}. If $p = 1$, the z_{A_j} signal will set flip-flop $(i+1)$, so that steps A_1 to A_j are not repeated. If, however, $p = 0$, the z_{A_j} signal sets flip-flop A_1, repeating steps A_1 to A_j. During the repetition, the z_{A_1} signal is produced through the AND gate having the output $y_{A_1} x_{A_j} y'_i$. These two decoding AND gates are necessary to avoid false actuation of z_{A_1}, should the x_i and x_{A_j} signals appear more than once during the cycle.

As with the $1/n$ code programmer solution, the number of

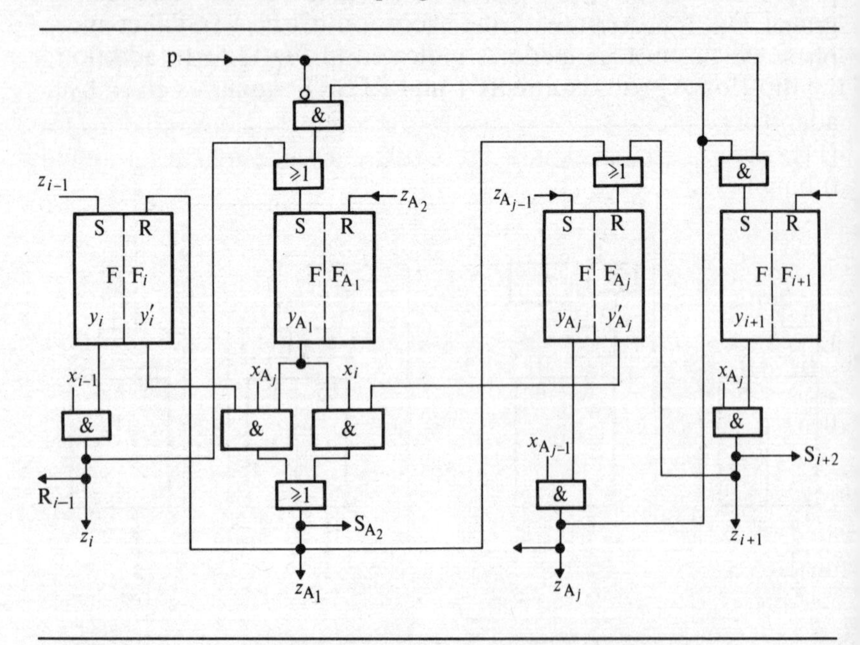

Fig. 5.9 $2/n$ code programmer for program with option of repeated steps (repeat for $p = 0$)

repeated steps must be at least three. The 2/*n* code programmer solution again requires seven additional gates, as compared to four additional gates for the 1/*n* code programmer solution.

Use of the Johnson or (1,2)/*n* codes seems to be impractical for this problem, because of the complicated SET and RESET connections required.

E. Conclusions

For multi-path programs with simultaneous parallel paths, the 1/*n*, 2/*n* and accumulating code programmers are all equally suitable, while the use of the Johnson or (1,2)/*n* codes is slightly more complicated.

For alternative parallel paths, for skipped steps, or for repeated steps, the 1/*n* code solution is the simplest, while the 2/*n* and accumulating code solutions require slightly more gates. The difference, however, is not sufficiently great to invalidate the conclusions arrived at in Chapter 3 (see Table 3.1).

The use of the Johnson and (1,2)/*n* codes for multi-path programs is more problematical, especially for the case of programs with repeated steps. The problem is due to the more complicated SET and RESET connections required by these codes.

It should be noted that programmer modules especially adapted for multi-path programs are commercially available (see [11]). Use of such modules can reduce the amount of additional tubing required.

Note

1. In cases where the switching speed of the flip-flops is unreliable, it might be safer to reset flip-flop *i* using an additional AND gate giving $R_i = y_{A_1} y_{B_1}$. This assures that the SET signal of flip-flop B_1 does not disappear before this flip-flop is completely set.

CHAPTER 6

PROGRAMMERS RECOGNIZING FALSE INPUT SIGNALS (ERROR DETECTING PROGRAMMERS)

A. Definition of the problem

In the programmers discussed so far, the completion of each program step is announced by an appropriate position-sensor signal fed back to the programmer as an input signal x_i. The programmer waits for a specific x_i signal corresponding to the present step, and will not react to changes in any x_i signals which do not pertain to the present step. The only exception to this principle so far appeared in connection with programs having alternative parallel paths (see Ch. 5). There, the programmer differentiates (but only at the branching point) between different input signals, and will produce an output signal depending on the input signal received.

The weakest elements in most sequential systems are usually the limit valves or limit switches which provide the x_i input signals. They are often exposed to shock and vibration, dirt, moisture, or corrosive atmosphere, and are usually the least reliable link in the control system. (The logic elements, by comparison, are usually protected by a separate enclosure.) In critical control systems, it may become desirable to have a continuous check on the proper operation of these limit valves or switches. In the case of false input signals (i.e., signals not corresponding to the desired program), various protective steps can be taken, regardless of whether the false input signal is caused by a defective position sensor or by a false cylinder motion.

Two questions must be discussed in connection with false input signals: the first is which input signals can (or should) be continuously monitored as to their correctness. The second question is how the programmer should react when a false input signal is discovered.

We shall first discuss the question of which input signals to monitor. Assume we have a normal programmer using any desired code, as described in the previous chapters, and at a certain program step i the cylinder B, for example, should extend. As the piston leaves its retracted position, the signal produced by position sensor $b-$ first changes sign; later, at the end of the stroke, $b+$ changes sign. The correct sequence followed by the two signals is $(b-,b+)$: (10),(00)(01).

The position sensor $b+$ could show faulty behaviour in either of two ways: it could neglect to produce a $b+$ signal at the end of the stroke, or it could produce a $b+=1$ signal before the completion of the stroke (for example, if the limit valve or switch should be jammed in its actuated position). In the first case, the program will be interrupted even with a normal programmer, since the (01) signal signifying the completion of the step is never obtained. In the second case, a normal programmer would simply skip the required step (falsely believing that the cylinder has already carried out its stroke), and the programmer would then continue with the program as if nothing had happened. To monitor this case, it is necessary to check that $b-$ changes from 1 to 0 while $b+$ is still 0.

In the discussion of this chapter, it becomes useful to differentiate between input signals that change during a given program step, and those input signals that do not change. In the above example, only $b-$ and $b+$ change during the step under discussion. A normal programmer will not react during this step if any input signal other than $b+$ gives a false signal (i.e., a signal not corresponding to the present cylinder positions according to the required program). Such false cylinder positions could occur, for example, if the program has been interrupted and then restarted, and if unknown events have taken place during the interruption. Such a situation could cause serious damage to the machine; for example, in machines where the paths of motion of different machine members cross each other. To prevent this, it would be necessary to monitor various position-sensor signals simultaneously, including those that do not normally change during that step, in order to make sure that these signals correspond with the required cylinder positions.

In the previous chapters, the 0 signals produced by unactuated position sensors were not utilized or taken into account. In this chapter, we shall sometimes take unactuated and actuated sensors into account on an equal basis. In other words, the programmer may react to certain 1 to 0 signal changes, just as it usually reacts only to certain 0 to 1 changes. Practically speaking, this means that the 0 signals emanating from unactuated position sensors must first be complemented before being sent to the programmer.

The second principal question to be discussed relates to the reaction of the programmer to a false input signal. The following three possibilities appear meaningful:

1. The programmer interrupts the program (possibly also actuating an alarm signal). After correction of the error, the program continues where interrupted.
2. The programmer does not interrupt the program, but only actuates an alarm signal.
3. The programmer is automatically reset to its starting position, and the program must be resumed with step No. 1.

We can elect to monitor the input signals during all program steps, or only during certain critical steps. Since the latter possibility is actually only a special case of the first one, these two possibilities need not be considered separately in the discussion to follow.

The three possibilities listed above will now be discussed in detail.

B. Program interrupt in case of false input signal

As already mentioned, we can differentiate between input signals that change during a given program step, and those that do not. We therefore have two cases to consider:

1. An input signal that must change during step i is false.
2. An input signal that should not change during step i is false.

In a normal programmer, case 1 produces either a program stop or a skipped step, while case 2 would have no effect at all.

In order to obtain a program stop for case 2, the SET input function of programmer stage i must be expanded. In the previous chapters, we used the symbol x_i to denote the programmer input signal which announces the completion of step i. We shall now define k_i as the Boolean AND product (conjunction) of *all* input signals existing at the completion of step i, including those input signals which do not change during this step. (In actual practice, k_i might include not all the input signals, but only those which are considered sufficiently important to be monitored during a given step.)

Figure 6.1 shows the resulting $1/n$ code programmer. Figure

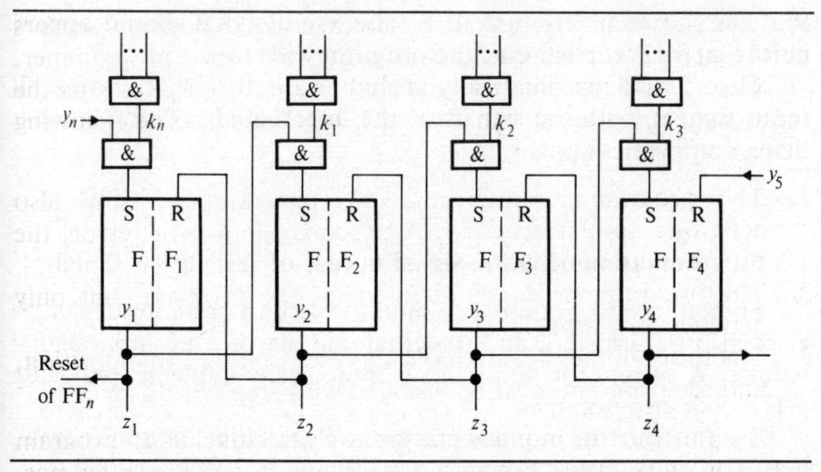

Fig. 6.1 $1/n$ code programmer providing program interrupt for false input signal

6.1 differs from Fig. 2.1 only by the additional AND gate at the SET input of each stage. If any false input signal, not corresponding to the required program, is included in k_i, the flip-flop cannot be set since k_i will be 0, and the program will come to a stop. This basic idea can be used with all codes treated in the previous chapters.

To take care of case 1, the AND gate providing the SET input for stage i must have the input x_i' in addition to y_{i-1} and

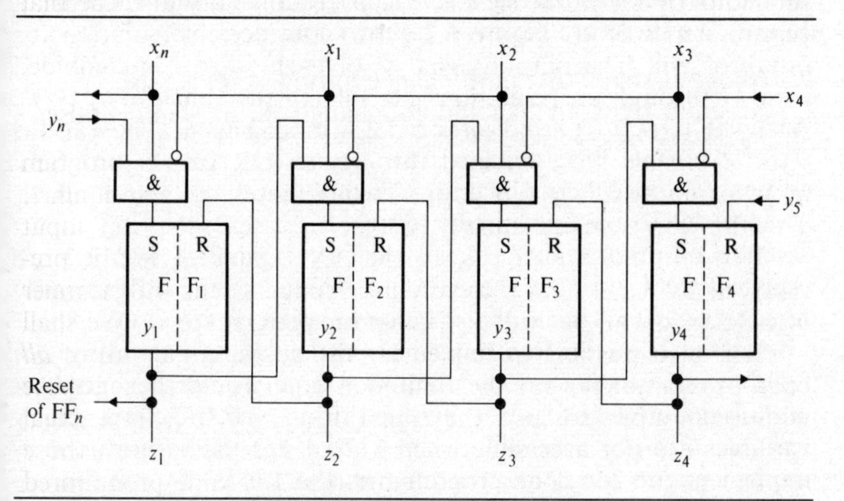

Fig. 6.2 $1/n$ code programmer providing program interrupt if false x_i signal appears before step i is carried out

x_{i-1}, as shown in Fig. 6.2. If a false $x_i = 1$ signal should appear before step i is carried out, the program will stop.

Case 2 will automatically include case 1 if k_i contains all input signals, with the signals of the unactuated position sensors being complemented.

C. Actuation of alarm signal in case of false input signal

Sometimes it is required to monitor certain or all input signals at each program step, and to actuate an alarm if an input signal is false. A system for doing this is more easily implemented using a $1/n$ code programmer.

To illustrate the implementation, we shall take, as an example, a system with three cylinders, A, B and C. We assume that, during step No. 3, cylinder B must extend its piston. During this step, we can only monitor the signals $a+$, $a-$, $c+$, $c-$ but not $b+$ or $b-$ since these two signals change during the step. (The signal $b+$ is only utilized at the completion of step No. 3, when the next flip-flop must be set.) The principle then is that, during a certain program step, we can monitor all input signals except those which change during that step.

The Boolean product (conjunction) of *all* input signals existing at the completion of step i was designated as k_i in the previous section. We shall now define the Boolean AND product (conjunction) of all input signals which remain constant during a certain step i as k_i^*. Figure 6.3 shows one possible solution for our problem. The output signal y_i of each stage i is combined with k_i^* through an Inhibition gate, giving the function $y_i \cdot (k_i^*)'$. A signal $y_i \cdot (k_i^*)' = 1$ indicates a false input signal. The various $y_i \cdot (k_i^*)'$ signals are combined through an OR gate to the SET input of an auxiliary flip-flop, which actuates the alarm signal. This flip-flop is reset manually (Q).

If the inputs $x_1, \ldots x_n$ to the AND gates of Fig. 6.3 are replaced by $k_1, \ldots k_n$, then a false input signal will not only actuate an alarm, but will also cause the program to stop.

It is also possible to implement the above circuit using $2/n$ code programmers, but the Inhibition gate would then need an additional input to give the signal $y_i \cdot y_{i+1} \cdot (k_i^*)'$. Since the y variables are not accessible when 3/2 or 5/2 valves are used to implement the $2/n$ code programmer, the $1/n$ code programmer would be preferable for this application whenever Category A logic elements are used.

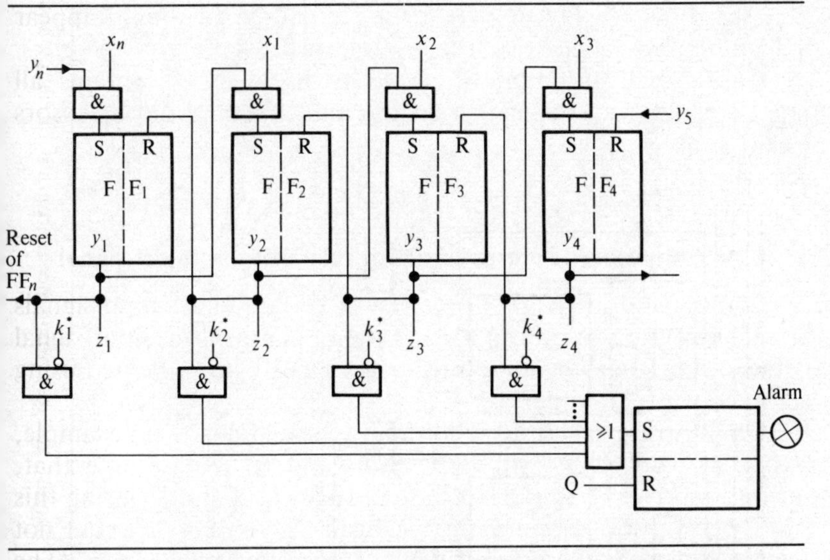

Fig. 6.3 $1/n$ code programmer actuating alarm for false input signal

Figure 6.4 illustrates a $1/n$ code programmer with the complete circuit required for monitoring all input signals, providing both an alarm signal and program interruption in case of false input signals. The sample program already used in Chapter 3 is used here too; i.e., a two-cylinder circuit with the required sequence START, A+, B+, B−, A−, B+, B−, STOP. To aid in the circuit design or for checking the circuit, it is advisable to prepare a table listing all input signals and the corresponding conjunctions k_i^* and k_i for every step, as shown below:

Step	Input signals $a-$	$a+$	$b-$	$b+$	k_i^*	k_i	x_i
0 = 6	1	0	1	0		$k_6=(a-)(a+)'(b-)(b+)'$	$x_6=b-$
0*a*	0	0	1	0	$k_1^*=(b-)(b+)'$		
1	0	1	1	0		$k_1=(a-)'(a+)(b-)(b+)'$	$x_1=a+$
1*a*	0	1	0	0	$k_2^*=(a-)'(a+)$		
2	0	1	0	1		$k_2=(a-)'(a+)(b-)'(b+)$	$x_2=b+$
2*a*	0	1	0	0	$k_3^*=(a-)'(a+)$		
3	0	1	1	0		$k_3=(a-)'(a+)(b-)(b+)'$	$x_3=b-$
3*a*	0	0	1	0	$k_4^*=(b-)(b+)'$		
4	1	0	1	0		$k_4=(a-)(a+)'(b-)(b+)'$	$x_4=a-$
4*a*	1	0	0	0	$k_5^*=(a-)(a+)'$		
5	1	0	0	1		$k_5=(a-)(a+)'(b-)'(b+)$	$x_5=b+$
5*a*	1	0	0	0	$k_6^*=(a-)(a+)'$		
6 = 0	1	0	1	0		$k_6=(a-)(a+)'(b-)(b+)'$	$x_6=b-$

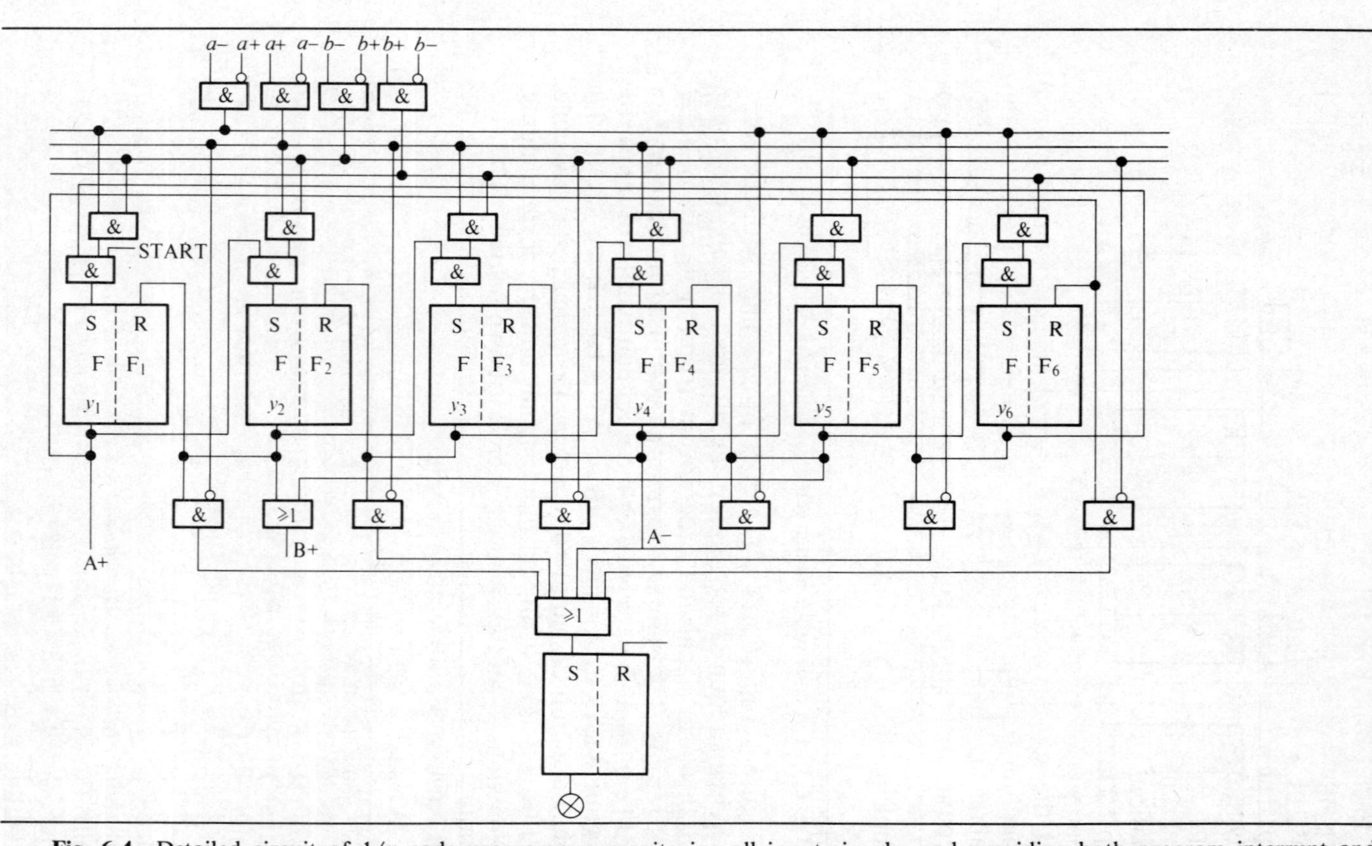

Fig. 6.4 Detailed circuit of $1/n$ code programmer monitoring all input signals, and providing both program interrupt and alarm for false input signal, for sample problem of Fig. 3.1

Each of the four Inhibition gates drawn at the top of Fig. 6.4 corresponds to a certain cylinder position. All four input signals must be correct ($k_i = 1$) in order to actuate the next program step. The auxiliary flip-flop at the bottom of the figure monitors only those input signals (k_i^*) not changing during a given program step.

From the above table, it is seen that there are really 12 combinations of input signals, even though there are only 6 program steps. This is so because, for each step, there are two changes of input signals: one when the cylinder leaves its initial position, and the second when the stroke is completed. Neither normal programmers nor the special programmers discussed in this chapter take the signal change at the beginning of the stroke into account. If it should be desired to monitor these signal changes too, a 12-stage programmer would be required for the above sample problem. Such a refinement, however, seems unnecessary, since each input signal is already monitored at the beginning and at the end of each step in the programmer of Fig. 6.4.

D. Resetting the programmer to starting position in case of false input signal

In some applications, there is no point in continuing the program once a false input signal has been discovered. In that case, the complete programmer must be reset. This basic idea, i.e., the resetting of the programmer in the case of an incorrect input-signal sequence, can serve as the basis for designing programmers for sequential systems with random inputs. This topic will be discussed in detail in Chapter 7.

We shall again use a $1/n$ code programmer. Regardless of the program step reached, the programmer is to be reset as soon as a false input signal is discovered. A possible solution is shown in Fig. 6.5. The circuit differs from a normal $1/n$ code programmer in that the flip-flop RESET inputs also include the $(k_i^*)'$ functions. Thus, the RESET function of flip-flop i is given by

$$\mathrm{R}_i = y_{i+1} + (k_i^*)'$$

where k_i^* is as defined previously. The signals produced by actuated limit valves or switches appear uncomplemented in k_i^*, while those of unactuated limit valves appear complemented. Since k_i^* must itself be complemented, it may be useful to differentiate between the signals of actuated and unactuated limit

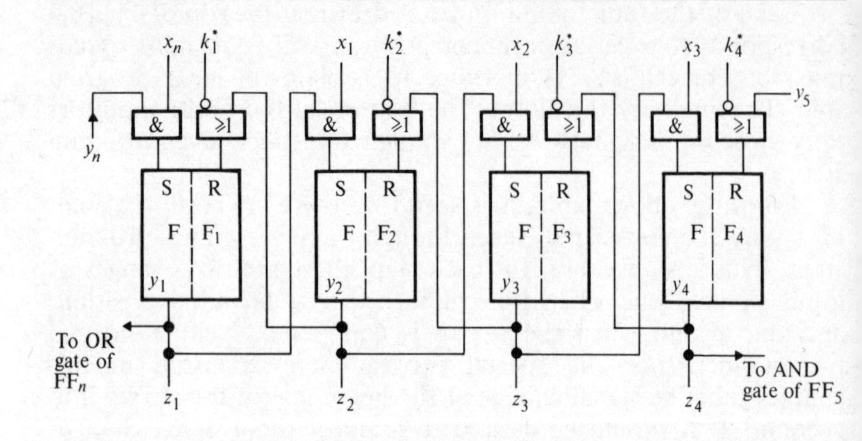

Fig. 6.5 $1/n$ code programmer with automatic programmer reset upon discovery of false input signal

valves. In Fig. 6.6, we assume that, during step No. 3 of a certain three-cylinder program, the motion C+ is to take place, motions A+ and B+ having taken place previously. Since the input signals $c+$ and $c-$ change value during this step, they are not included in k_3^*. Figure 6.6(a) only includes the input signals produced by the actuated limit valves, so that $a+$ and $b+$ are used as inputs of an AND gate. In Fig. 6.6(b), only the input signals

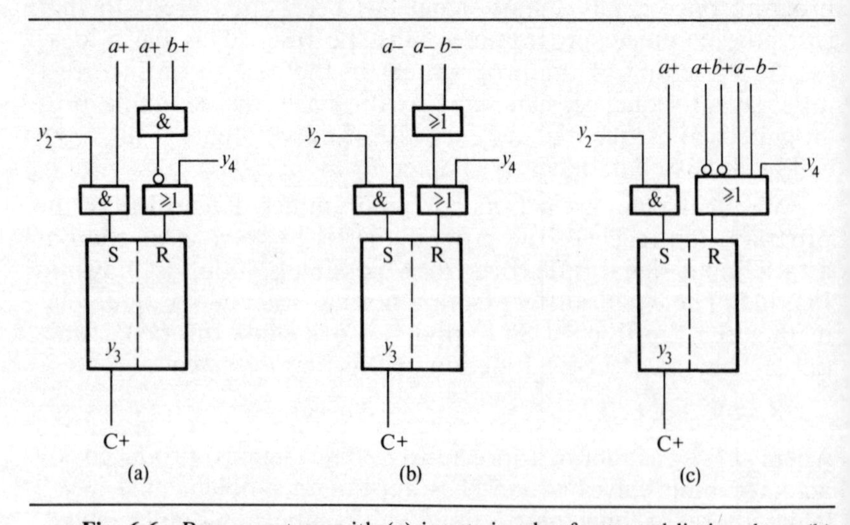

Fig. 6.6 Program stage with (a) input signals of actuated limit valves, (b) input signals of unactuated limit valves, (c) input signals of both actuated and unactuated limit valves

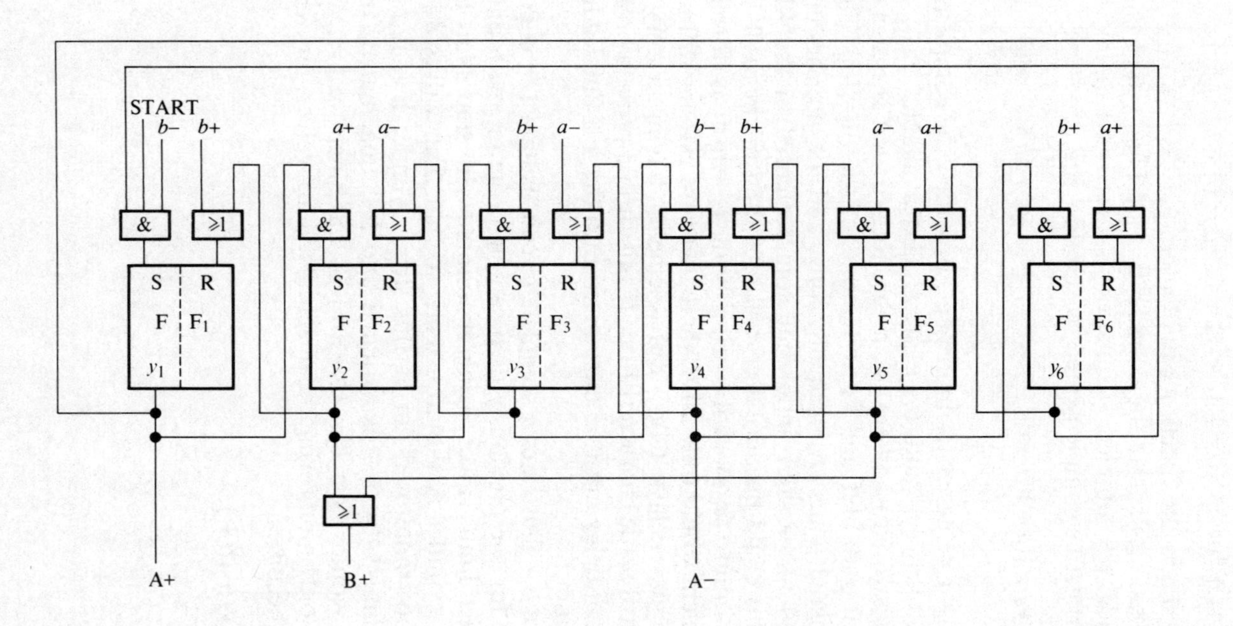

Fig. 6.7 Detailed circuit of $1/n$ code programmer with automatic programmer reset upon discovery of false input signal of unactuated limit valves; sample problem of Fig. 3.1

$a-$ and $b-$ of the unactuated limit valves are considered. These signals must therefore be complemented, and then fed into an AND gate whose output itself must be complemented. Using De Morgan's theorem, we can write

$$R_3 = y_4 + [(a-)'(b-)']' = y_4 + (a-) + (b-)$$

so that a single OR function can be used.

Figure 6.6(c) shows the combination of Figs. 6.6(a) and (b), and includes the complete k_3^* function $k_3^* = (a+)(b+)(a-)'(b-)'$. Since k_3^* must be complemented, we could write

$$R_3 = y_4 + [(a+)(b+)(a-)'(b-)']' = y_4 + (a+)' + (b+)' + (a-) + (b-)$$

which might be easier to implement, provided the complemented input signals $(a+)'$ and $(b+)'$ are already available.

In Figs. 6.6(a) and (c) it can be noticed that the signal $a+$ appears in both SET and RESET inputs of the flip-flop. The respective signal paths are unequally long, and $a+$ appears uncomplemented in the SET input but complemented in the RESET input. This could produce a critical race, and might result in the flip-flop inputs temporarily being both 1, which is forbidden. The problem would only arise where there are extreme differences in the signal delays (e.g., with elements having a very non-symmetrical static characteristics). If this should prove a problem, an artificial delay must be introduced into the $a+$ line leading to the SET input.

Figure 6.7 illustrates a complete circuit using a $1/n$ code programmer for the previously used sample program, with only the unactuated limit valves being monitored. In step No. 1, for example, cylinder A should extend, so that limit valves $a+$ and $a-$ are not monitored during this step. Cylinder B is retracted, so that limit valve $b-$ is actuated and $b+$ unactuated. In this case, we would have $k_1^* = (b-)(b+)'$, but since we are, in this example, only monitoring the unactuated limit valves, our RESET function becomes

$$R_1 = y_2 + (b+)$$

CHAPTER 7

PROGRAMMERS FOR RANDOM INPUT SIGNALS

A. Definition of the problem

Whereas the programmers discussed in Chapters 1 to 4 dealt with programs having a fixed sequence of steps (i.e., repeating cycles), this chapter deals with programmers whose input signals can change in a random manner. However, it is assumed throughout that only a single input variable can change at any given moment. The programmer is to produce certain output signals only if the input signals change in a certain predefined order. The programmers described in Chapter 6 were a step in this direction. Those programmers, although designed to produce a certain program cycle, were able to recognize deviations from the desired normal cycle. In the programmers discussed in this chapter, the input signals are completely random; i.e., the programmer output signals generally have no effect on the succeeding input signals.

We shall now define the term *input word* as a sequence of input states. Suppose a system has three simultaneous input signals a_1, a_2 and a_3 respectively. If, for example, these input signals have the values $a_1 = 0$, $a_2 = 1$, $a_3 = 1$, and if a_3 then changes from 1 to 0, we would have the following input word:

$(a_1 a_2 a_3)$: (011),(010)

In similar fashion, the sequence of output signals $z_1 \ldots z_n$ of a system can be expressed as an *output word*.

In the programmers described in this chapter, a certain output word is produced only provided the programmer senses the required input word. With some problems, the relation between the input word and the resulting output word is easily formulated; if so, the method described here is especially suitable, and will generally lead to a circuit design with less effort than the conventional method described in [1]. In other applications, formulating the problem in terms of input and output words is more difficult.

The basic idea used in designing programmers for random input signals is the same as that used in the previous chapter: as long as the input states follow a certain desired sequence, each programmer stage is activated in turn. If, however, a 'false' input state appears, i.e., one deviating from the required order, then all programmer stages are reset, and the programmer must begin anew from the starting position. This basic idea is generally used with problems described in this chapter. However, in some cases it may not be convenient for all programmer stages to reset the register if a false input state appears. In such cases, a combination of programmer stages described in this chapter and those of previous chapters is possible.

Sometimes certain input words appear repeatedly in a complex problem. Each such input word can then be handled by a subprogrammer, and the various subprogrammers combined with appropriate output networks. Sometimes two or more different input words are partially identical, and this fact can be utilized to simplify the design. The resulting system structures have some similarity with the circuits described in Chapter 5 (Programmers for multi-path programs). For a more detailed discussion of these points, see [17].

B. Formulating problems using input and output words

To define or formulate any given problem, it is necessary to determine the required input word which will cause each output variable to become 1 (or possibly 0). Such input words can then be used to directly calculate the SET and RESET functions of the various programmer flip-flops, using equations to be presented later. Furthermore, the output variables can be defined as functions of the various flip-flop outputs.

In cases where individual logic gates (e.g., elements of Category B and C) are used to construct feedback flip-flop circuits (see, e.g., Figs. 3.9 and 3.12), it is usually more advantageous to express the solution in the form of memory excitation functions, rather than flip-flop SET and RESET functions. The necessary equations for this case will be presented in Section D of this chapter.

The determination of the relation between input and output words will now be demonstrated by means of three different examples. The first one was chosen as a very simple example where no encoding is necessary. The second example presents a problem having a single output variable, but where the required input word is fairly long. The third problem was chosen to illustrate a case with two output variables, and where the required input word can be shortened, provided additional starting conditions are assumed.

Example 7.1

The direction of rotation of a shaft is to be determined, using appropriate sensors and switching system. For a clockwise rotation, output variable $z = 1$ should result, whereas $z = 0$ indicates counter-clockwise rotation.

Figure 7.1 shows a possible sensor arrangement, using a cam attached to the rotating shaft, and two sensors spaced 90 degrees apart. The two sensors supply the signals a_1 and a_2 respectively whenever they are covered by the cam. For a clockwise rotation, the input word (assuming the starting position shown in the figure) will be

(a_1a_2): (00),(10),(11),(01)

Any two successive input states are sufficient to determine the direction of rotation. For example, the input word (00),(10) could be used to produce a SET signal for a flip-flop; i.e., S = 1 only if the input state (00) appears first and (10) afterwards. This can be written as follows:

(a_1a_2): (00),(10)
S: 0 , 1

For a counterclockwise rotation, the input word, which must

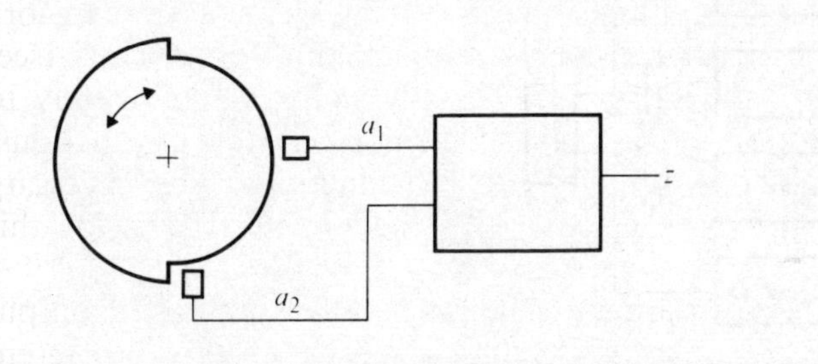

Fig. 7.1 Sensor arrangement for determining direction of shaft rotation (Example 7.1)

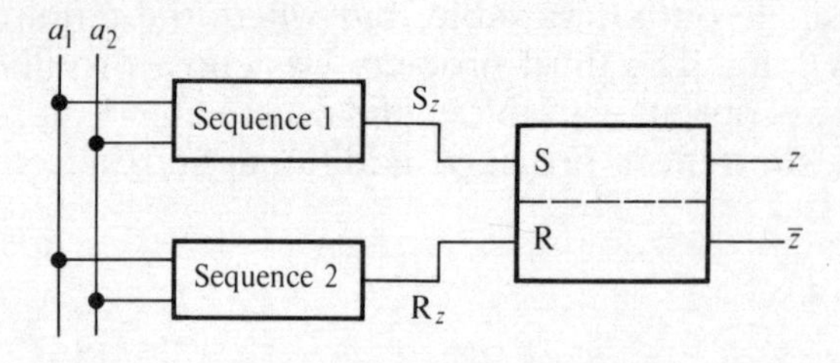

Fig. 7.2 Switching system for arrangement of Fig. 7.1

reset the flip-flop, will be:

$$
\begin{array}{rl}
(a_1a_2): & (10),(00) \\
\text{R}: & 0\;,\;1
\end{array}
$$

Up to one complete shaft rotation may here be required to discover a change in the direction of rotation, depending on the cam position when the change takes place. Figure 7.2 shows the required switching system schematically.

A different solution to this problem would utilize all four two-state input words appearing during one clockwise rotation. Each such two-state input word would activate an auxiliary switching circuit, with each such circuit supplying a 1 output signal during one-quarter rotation. Connecting these four output signals through an OR gate, we obtain a continuous gate output during the clockwise rotation. A change in direction of rotation will already make itself felt after a quarter rotation. The various input and output words are given below, and the required switching system is shown schematically in Fig. 7.3.

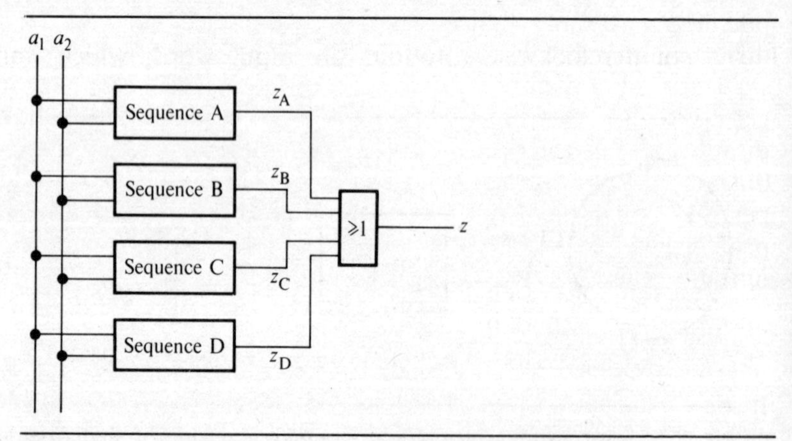

Fig. 7.3 Alternative switching system for arrangement of Fig. 7.1

(a_1a_2): (00),(10); (10),(11); (11),(01); (01),(00)
z_A: 0 , 1 ; z_B: 0 , 1 ; z_C: 0 , 1 ; z_D: 0 , 1
$z = z_A + z_B + z_C + z_D$

In example 7.1, only two-state input words are used. An example using much longer input sequences is that of a sequential-action lock.

Example 7.2

A lock with electric actuation has three input toggle switches. The lock will only open if these switches are actuated in a certain specified sequence. The two positions of each toggle switch are defined as 1 and 0 respectively. The lock opens ($z = 1$) only if the switches are actuated according to the following input word:

$(a_1a_2a_3)$: (001),(011),(010),(110),(100),(101),(100)
z: 0 , 0 , 0 , 0 , 0 , 0 , 1

Example 7.3

A switching system for supplying a warning signal has two inputs: One is connected to the signal to be monitored for disturbances, with a 1 signal signifying a disturbance and a 0 signal no disturbance. The second input is connected to an 'Acknowledgement' button, which is pressed by the operator to signify that the warning signal has been noted. The system has two outputs: $z_1 = 1$ represents an unacknowledged warning signal (possibly a siren or buzzer), whereas $z_2 = 1$ represents an acknowledged warning possibly a warning light) which is maintained as long as the disturbance still persists.

If $z_1 = z_2 = 0$ and a disturbance appears ($a_1 = 1$), then we must get $z_1 = 1$, and this signal remains even if the disturbance should disappear. Only after the button is pressed ($a_2 = 1$) does z_1 become 0 again, but z_2 will then become 1 provided the disturbance still persists. When the disturbance disappears, z_2 becomes 0 also. The problem can be expressed by means of the following input and output words:

(a_1a_2): (00),(10),(00) (a_1a_2): (11),(10)
z_1: 0 , 1 , 1 z_2: 1 , 1

It should be noted that any sequence different from the above will cause z_1 or z_2 respectively to be 0. If, for example, a_2 should become 1 at the end of the above input word for z_1, then we will get $z_1 = 0$, since the state (01) is not part of the above-listed required input word.

It is desirable to keep the necessary input words as short as possible, since this saves flip-flops. In the above example, for instance, the input word for z_1 could be shortened to

$$\begin{array}{ll} (a_1a_2): & (10),(00) \\ z_1: & \;\;1\;\;,\;\;1 \end{array}$$

under the condition that, for $(a_1a_2) = (10)$, we have $z_2 = 0$. This condition is necessary since, with the shortened input word sequence, we might get $z_1 = 1$ and $z_2 = 1$ at the same time. Since the disturbance should either appear as being acknowledged or as being unacknowledged but not both, we do not wish to have $z_1 = z_2 = 1$.

C. Design of 1/*n* code programmers using set–reset flip-flops

Formulating the problem in terms of input and output words is a prerequisite for the programmer design. The next step consists of encoding the individual input states. In theory, any one of the codes utilized in previous chapters could be used here. However, the $1/n$ code results in the simplest equations, and will therefore be used here.

The case where the various programmer stages consist of set-reset flip-flops will be treated in this section, with our aim being to calculate the necessary SET and RESET functions.

The encoding of the input word consists of assigning each input state to a programmer flip-flop output variable (or to a combination of variables, in the case of codes other than $1/n$ code). However, in contrast to the programmers discussed in the previous chapters, the course of events is now non-cyclic. It is possible to return to the first step from anywhere in the input-state sequence, but first all of the programmer stages must be reset. This fact, however, need not influence the encoding process, which is illustrated below for the problem of Example 7.2:

If an input word consists of only two input states, no encoding at all is necessary. A single flip-flop would then be sufficient, and this is set once the first input state of the input word has appeared. This flip-flop will also remain set during the second input state, and the system will differentiate between the two states only by means of the different input signals.

The SET- and RESET-functions of the various flip-flops are now calculated by means of the following equations, which are

$i =$	1	2	3	4	5	6	7
$(a_1a_2a_3)$:	(001),	(011),	(010),	(110),	(100),	(101),	(100)
z:	0 ,	0 ,	0 ,	0 ,	0 ,	0 ,	1
y_1:	1	0	0	0	0	0	0
y_2:	0	1	0	0	0	0	0
y_3:	0	0	1	0	0	0	0
y_4:	0	0	0	1	0	0	0
y_5:	0	0	0	0	1	0	0
y_6:	0	0	0	0	0	1	0
y_7:	0	0	0	0	0	0	1

derived using similar ideas as those used in Section D of Chapter 6.

$$S_1 = k_1 \quad (7\text{-}1)$$
$$S_i = x_{i-1}y_{i-1} \qquad (i \neq 1) \quad (7\text{-}2)$$
$$R_i = (k_i + k_{i+1})' + y_{i+1} \quad (7\text{-}3)$$

$(k_i + k_{i+1})'$ in equation (7-3) should be skipped, if, for certain stages, the programmer is to act as described in previous chapters (no programmer-reset, if false input state). The symbol k_i here has the same meaning as in Chapter 6; i.e., represents the Boolean AND product (conjunction) of all input-variable signals present during step or state i. For the Example 7.2, for instance, the first input state is (001), so that $k_1 = a_1'a_2'a_3$.

In programmers for cyclic programs, the symbol x_i was defined as the feedback signal announcing completion of step i. In an analogous fashion, we shall here define x_{i-1} in equation (7-2) as that input variable which changes value between input states $i - 1$ and i of the input word. The variable assumes that value corresponding to the new input state i. For example, if we wish to calculate S_2 for the above Example 7.2, we would have $i = 2$, and equation (7-2) would give

$$S_2 = x_1y_1 = a_2y_1$$

since a_2 is the variable changing between input states 1 and 2, and since a_2 assumes the value 1 during input state 2. (The term x_{i-1} in equation (7-2) could be changed to k_i to prevent malfunctioning in case two input signals change almost simultaneously, even though we have assumed at the beginning that such simultaneous changes are not permitted.)

It should be pointed out that the k_{i+1} term in equation (7-3) is required to avoid hazards; i.e., to prevent flip-flop i from being reset before flip-flop $i + 1$ is completely set.

The application of equations (7-1) to (7-3) will now be illustrated, using the three examples previously discussed.

Example 7.1 – first formulation

We refer to Fig. 7.2, and to the input and output words already listed in the previous section, but repeated here:

$$(a_1a_2):\quad (00),(10) \qquad (a_1a_2):\quad (10),(00)$$
$$S_z:\quad 0\ ,\ 1 \qquad R_z:\quad 0\ ,\ 1$$

Both words consist of only two input states each, so that no encoding is required. Thus, for each word, a single flip-flop is sufficient, which has to be set with the appearance of the first input state of the word, and which also remains set during the second input state. Assigning the index S for the two inputs of the flip-flop needed for the S_z output function, and the index R for the two inputs of the second flip-flop required for R_z, we get the equations:

$$S_S = k_1 = a_1'a_2'$$
$$R_S = (k_1 + k_2)' + y_2 = (a_1'a_2' + a_1a_2')' + 0 = a_2$$
$$S_R = k_1 = a_1a_2'$$
$$R_R = (k_1 + k_2)' + y_2 = (a_1a_2' + a_1'a_2')' + 0 = a_2$$

Since both flip-flops will remain set during the second input state of either input word, we must use decoding to get output signals S_z and R_z only for the second state of each of the two words. For decoding purposes, it is best here to use that variable which changes value between the first and second input states, which would be a_1 (changing from 0 to 1) for the S_z input word, and also a_1 (but changing from 1 to 0) for the R_z input word. The final output functions thus become

$$S_z = a_1y_S \qquad R_z = a_1'y_R$$

The complete solution, using three flip-flops, is shown in Fig. 7.4.

Example 7.1 – second formulation

In the previous section, a second formulation for Example 7.1 was presented, using four input words of two states each. A solution based on this formulation will, of course, use more flip-flops, but has the advantage that a change in direction of rotation is recognized after no more than one quarter rotation.

Applying the procedure demonstrated above to the four input words already listed, we obtain the following equations. (Note that S_A and R_A are identical to S_S and R_S of the previous formulation, since the respective input words are the same.)

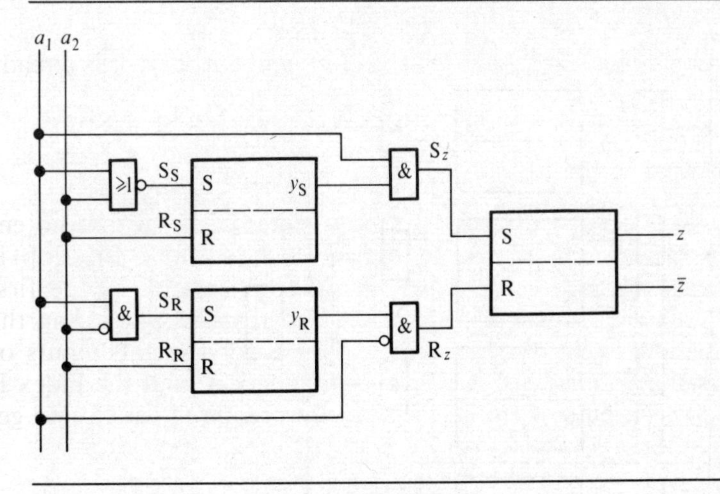

Fig. 7.4 Complete solution corresponding to Fig. 7.2

$$S_A = S_S = a_1'a_2' \qquad z_A = a_1y_A$$
$$R_A = R_S = a_2$$
$$S_B = a_1a_2' \qquad z_B = a_2y_B$$
$$R_B = (a_1a_2' + a_1a_2)' = a_1'$$
$$S_C = a_1a_2 \qquad z_C = a_1'y_C$$
$$R_C = (a_1a_2 + a_1'a_2)' = a_2'$$
$$S_D = a_1'a_2 \qquad z_D = a_2'y_D$$
$$R_D = (a_1'a_2 + a_1'a_2')' = a_1$$
$$z = z_A + z_B + z_C + z_D$$

Figure 7.5 shows the complete solution, which requires four flip-flops.

Example 7.2

Referring to the encoding table for this problem already presented at the beginning of this section, application of equations (7-1) to (7-3) yield:

$$S_1 = k_1 = a_1'a_2'a_3$$
$$R_1 = (k_1 + k_2)' + y_2 = (a_1'a_2'a_3 + a_1'a_2a_3)' = a_1 + a_3' + y_2$$
$$S_2 = x_1y_1 = a_2y_1$$
$$R_2 = (k_2 + k_3)' + y_3 = (a_1'a_2a_3 + a_1'a_2a_3')' + y_3 = a_1 + a_2' + y_3$$

•

•

•

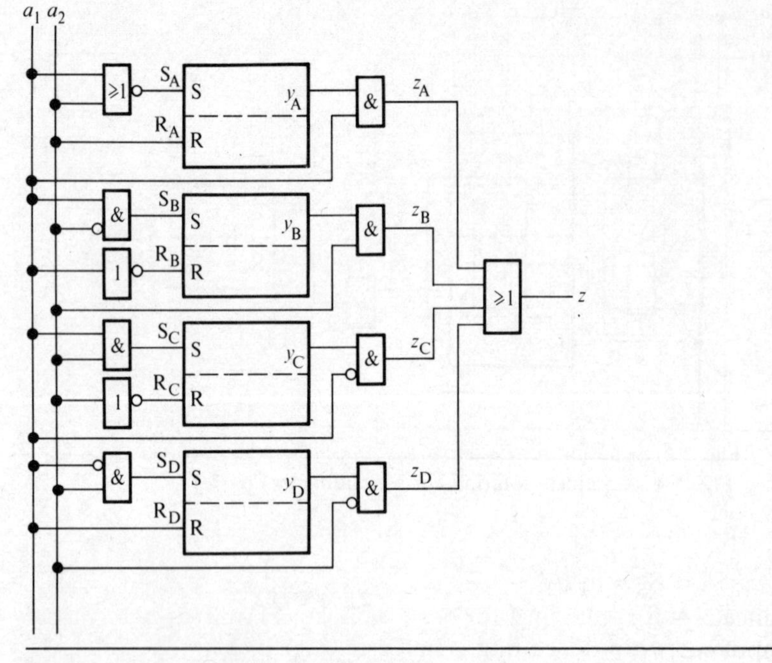

Fig. 7.5 Complete solution corresponding to Fig. 7.3

•

•

$$S_7 = x_6 y_6 = a_3{}' y_6$$
$$R_7 = (k_7 + k_8)' + y_8 = (a_1 a_2{}' a_3{}' + 0)' + 0 = a_1{}' + a_2 + a_3$$
$$z = y_7$$

The resulting solution is shown in Fig. 7.6.

Example 7.3

In order to save flip-flops, we shall use here the previously listed shortened input word for z_1, together with the condition that $z_2 = 0$ for $(a_1 a_2) = (10)$. We thus have

$(a_1 a_2)$:	(10), (00)	$(a_1 a_2)$:	(11), (10)
z_1:	1 , 1	z_2:	1 , 1

Assumption: $(z_2 = 0)$

The above assumption is taken into account by using the AND product of $z_2{}'$ and k_1 for setting flip-flop 1. The equations are

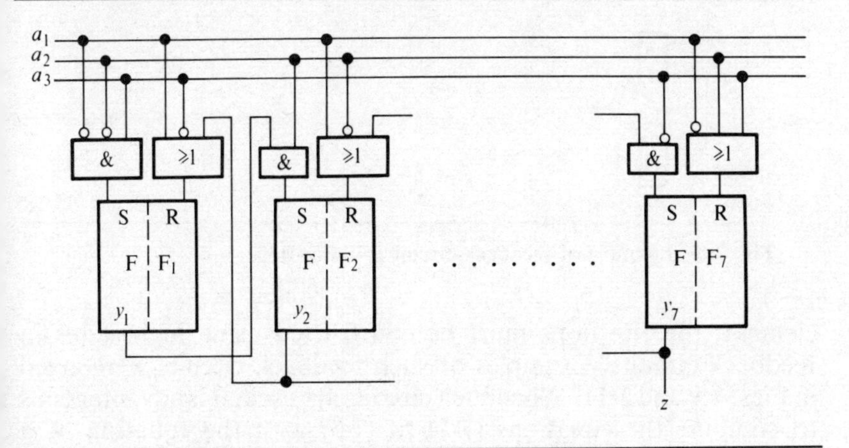

Fig. 7.6 Solution for Example 7.2

$$S_1 = k_1 z_2' = a_1 a_2' z_2' \qquad S_2 = k_1 = a_1 a_2$$
$$R_1 = (k_1 + k_2)' \qquad R_2 = (k_1 + k_2)'$$
$$= (a_1 a_2' + a_1' a_2')' = a_2 \qquad = (a_1 a_2 + a_1 a_2')' = a_1'$$

Since z_1 and z_2 remain 1 for both input states of their two respective input words, no decoding is necessary, so that $z_1 = y_1$ and $z_2 = y_2$. The complete solution is shown in Fig. 7.7.

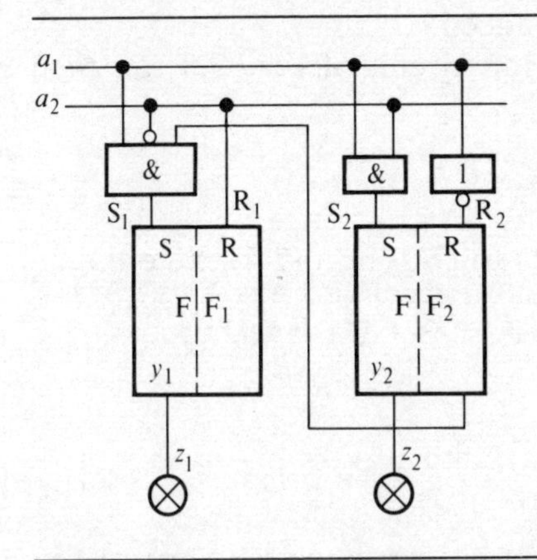

Fig. 7.7 Solution for Example 7.3

D. Design of 1/*n* code programmers using feedback-circuit flip-flops

When using element systems possessing no set-reset flip-flop

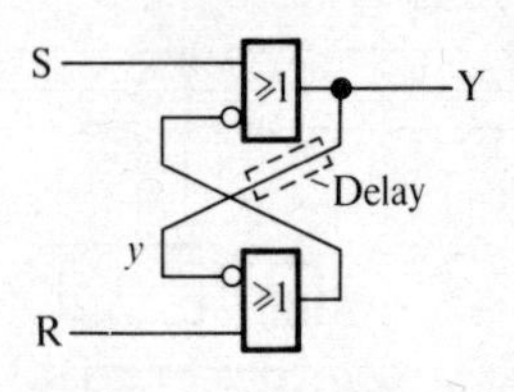

Fig. 7.8 Example of feedback-circuit R–S flip-flop

element, the flip-flops must be constructed using logic gates in feedback circuits. Examples of such feedback circuits were given in Figs. 3.9 and 3.11. When such circuits are used, it is advantageous to combine the equations (7-1) to (7-3) with the equation of a set-reset flip-flop.

Figure 7.8 shows a set-reset flip-flop constructed out of two Implication gates connected in a feedback circuit, identical to that of Fig. 3.11. If we wish to write the equation of this feedback circuit, we must differentiate between the output variable Y and the feedback variable y. Usually, the ys are referred to as secondaries, and the Ys as the excitations. We can visualize the two variables Y and y as separated by a delay as shown in the figure (e.g., the signal transfer time), although, practically speaking, no separate delay element is introduced.

The equation of the flip-flop circuit of Fig. 7.8 can then be written as

$$Y = S + (R + y')' = S + R'y \tag{7-4}$$

Substituting the equations (7-1) to (7-3) into (7-4), we obtain

$$Y_1 = k_1 + [(k_1 + k_2)' + y_2]'y_1 = k_1 + (k_1 + k_2)y_2'y_1 \tag{7-5}$$

and, for $i \neq 1$

$$\begin{aligned} Y_i &= x_{i-1}y_{i-1} + [(k_i + k_{i+1})' + y_{i+1}]'y_i = \\ &= x_{i-1}y_{i-1} + (k_i + k_{i+1})y_{i+1}'y_i \end{aligned} \tag{7-6}$$

For the special case of an input word consisting of two input states only so that no decoding is required, we have $y_2 = 0$, so that equation (7-5) is simplified to

$$Y = k_1 + (k_1 + k_2)y \tag{7-7}$$

Application of the above equations will be briefly illustrated using some of the examples of the previous section.

Example 7.1 – first formulation

(a_1a_2):	(00), (10)	(a_1a_2):	(10), (00)
S_z:	0 , 1	R_z:	0 , 1

Both input words consist of only two input states, so that no encoding is required and equation (7-7) can be used.

For the S_z function, $k_1 = a_1'a_2'$ and $k_2 = a_1a_2'$, while for R_z we have $k_1 = a_1a_2'$ and $k_2 = a_1'a_2'$. Substituting into equation (7-7), we get the following two memory-circuit excitation equations:

$$Y_S = a_1'a_2' + (a_1'a_2' + a_1a_2')y_S = a_2'(a_1' + y_S)$$
$$Y_R = a_1a_2' + (a_1a_2' + a_1'a_2')y_R = a_2'(a_1 + y_R)$$

Decoding is necessary, since both S_z and R_z should only be 1 during the second input state of their respective input words. Since, in the S_z input word, the variable a_1 changes from 0 to 1, we use a_1 for decoding, and write

$$S_z = a_1y_S$$

Similarly, in the R_z input word, a_1 changes from 1 to 0, and we therefore write

$$R_z = a_1'y_R$$

Figure 7.9 shows the complete solution. (The set-reset flip-flop shown in the figure could, in this case, be constructed using a circuit such as that of Fig. 7.8.)

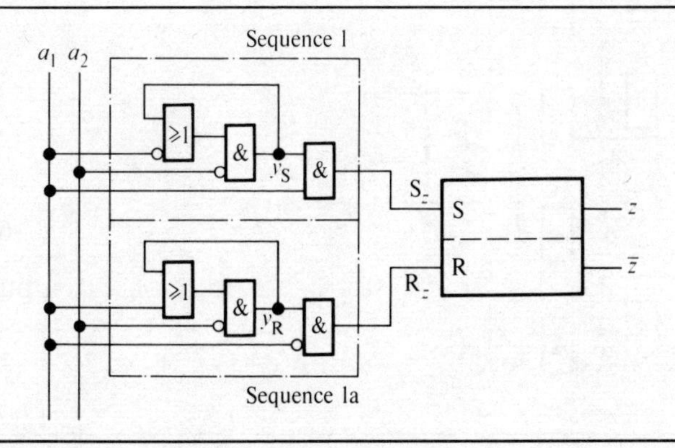

Fig. 7.9 Solution of Example 7.1 (First Formulation) using feedback memory circuits

Example 7.1 – second formulation

The input and output words were formulated in Section C as

(a_1a_2): (00),(10); (10),(11); (11),(01); (01),(00)
z_A: 0 , 1 ;z_B: 0 , 1 ;z_C: 0 , 1 ;z_D: 0 , 1

$$z = z_A + z_B + z_C + z_D$$

As before, the output function z_A is determined by the same input word as S_z in the previous formulation, so that $z_A = S_z$. Using the same indices for the excitation and secondary variables as for the output functions, we get, using equation (7-7), the following relations:

$$Y_A = a_1'a_2' + (a_1'a_2' + a_1a_2')y_A = a_2'(a_1' + y_A)$$
$$Y_B = a_1a_2' + (a_1a_2' + a_1a_2)y_B = a_1(a_2' + y_B)$$
$$Y_C = a_1a_2 + (a_1a_2 + a_1'a_2)y_C = a_2(a_1 + y_C)$$
$$Y_D = a_1'a_2 + (a_1'a_2 + a_1'a_2')y_D = a_1'(a_2 + y_D)$$
$$z_A = a_1y_A$$
$$z_B = a_2y_B$$
$$z_C = a_1'y_C$$
$$z_D = a_2'y_D$$
$$z = z_A + z_B + z_C + z_D$$

Figure 7.10 shows the complete solution. If we had attempted to construct the circuit of Fig. 7.5 using individual logic elements,

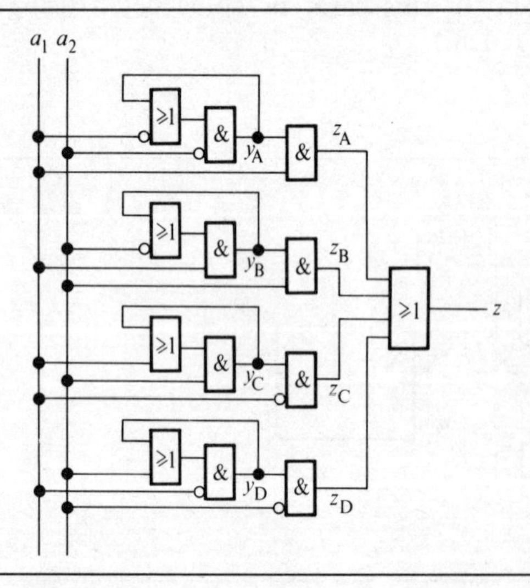

Fig. 7.10 Solution of Example 7.1 (Second Formulation) using feedback memory circuits

we would need two elements per flip-flop, in addition to the other gates shown. It is seen from Fig. 7.10 that the use of equation (7-7) saves six elements in this case.

Example 7.2

Using the encoding table presented in the previous section, we use equation (7-5) to calculate Y_1 and equation (7-6) for the remaining excitation equations. In calculating Y_7, we set $y_8 = 0$. The results are:

$$Y_1 = k_1 + (k_1 + k_2)y_2'y_1 = a_1'a_2'a_3 + (a_1'a_2'a_3 + a_1'a_2a_3)y_2'y_1 = a_1'a_2'a_3 + a_1'a_3y_2'y_1$$
$$Y_2 = x_1y_1 + (k_2 + k_3)y_3'y_2 = a_2y_1 + (a_1'a_3 + a_1'a_2a_3')y_3'y_2 = a_2y_1 + a_1'a_2y_3'y_2$$
$$Y_3 = x_2y_2 + (k_3 + k_4)y_4'y_3 = a_3'y_2 + (a_1'a_2a_3' + a_1a_2a_3')y_4'y_3 = a_3'y_2 + a_2a_3'y_4'y_3$$
$$Y_4 = x_3y_3 + (k_4 + k_5)y_5'y_4 = a_1y_3 + (a_1a_2a_3' + a_1a_2'a_3')y_5'y_4 = a_1y_3 + a_1a_3'y_5'y_4$$
$$Y_5 = x_4y_4 + (k_5 + k_6)y_6'y_5 = a_2'y_4 + (a_1a_2'a_3' + a_1a_2'a_3)y_6'y_5 = a_2'y_4 + a_1a_2'y_6'y_5$$
$$Y_6 = x_5y_5 + (k_6 + k_7)y_7'y_6 = a_3y_5 + (a_1a_2'a_3 + a_1a_2'a_3')y_7'y_6 = a_3y_5 + a_1a_2'y_7'y_6$$
$$Y_7 = x_6y_6 + (k_7 + k_8)y_8'y_7 = a_3'y_6 + (a_1a_2'a_3' + 0)0'y_7 = a_3'y_6 + a_1a_2'a_3'y_7$$
$$z = y_7$$

In some of the above equations, x_i could be factored out. Since this is not possible with all the equations, it has not been done anywhere, in order to obtain an identical structure for all the feedback circuits.

CHAPTER 8

PROGRAMMERS NOT BASED ON LOGIC COUNTERS

The purpose of this chapter is to discuss briefly other types of sequential programmers, and compare them to the logic-counter programmer that is the main subject of this book. These other types of programmers are either based on mechanical devices to store the required program sequence, or on microcomputers using software programming. The following types will be discussed here:

A. Mechanical Programmers
   1. Card readers or tape readers
   2. Stepping drums
   3. Matrix boards

B. Software Programmers
   4. Programmable logic controllers (P.L.C.)
   5. Microprocessors

A. Mechanical Programmers

1. Card or tape readers

These exist in both electric and pneumatic versions. The program sequence is stored in the card or tape by punching holes in the appropriate tracks. The number of external devices that can be actuated depends on the number of parallel tracks available; i.e., on the width of the card or tape. The number of program steps that can be accommodated is limited by the card length. Tape readers have the obvious advantage over card readers that the tape length is essentially unlimited. Also, tapes can easily be formed into a closed loop, so that the program will automatically repeat itself when completed.

In electric readers, the holes are 'read' either by means of photoelectric devices, or by metal brushes contacting a conducting plate on the other side of the card or tape. The brush must be wide enough to provide signal continuity for consecutive holes ('make-before-break'). A typical punched tape controller [19], uses brushes having a 1/4 ampere rating, and is able to accommodate from 12 to 82 channels, depending on tape reader size.

Another card reader [20], markets a program card controller using a ribbed-track plastic card, for programming 10, 22 or 48 circuits. The tracks actuate limit switches mechanically, and the track must be cut away wherever a switch action is desired. By actuating switches directly, loads up to 10 amperes can be switched.

The card or tape drive can either be synchronous or event-controlled. With synchronous drives, the card or tape is advanced by a synchronous motor at constant speed, and the system, in effect, is nothing more than a multi-contact timer. With event-controlled drives, a stepping motor (or equivalent indexing arrangement) is used to advance the tape or card to the next row whenever a command pulse is received, indicating that the previous program step has been completed.

Pneumatic card or tape readers generally detect the presence of a hole by means of the circuit shown in Fig. 8.1. The principle is similar to that of a flapper-nozzle system or back-pressure sensor, with the card or tape taking the place of the flapper. Presence of a hole permits the air to escape, so that the pressure in the corresponding line assumes the Logic 0 value. Absence of a hole produces Logic 1.

A programmer of this type, see [21], is intended especially for use with fluidic elements. Metal cards are used, with the punching done on a special punching jig. The card can accommodate up to 20 circuits and 40 program steps. The punched card is wound on a drum which is indexed mechanically through an angle corresponding to one row each time a 0.7 bar (10 psi) signal is received. The programmer is able to supply 0.35–0.7 bar (5–10 psi) output signals.

Another card reader programmer, see [22], uses standard IBM

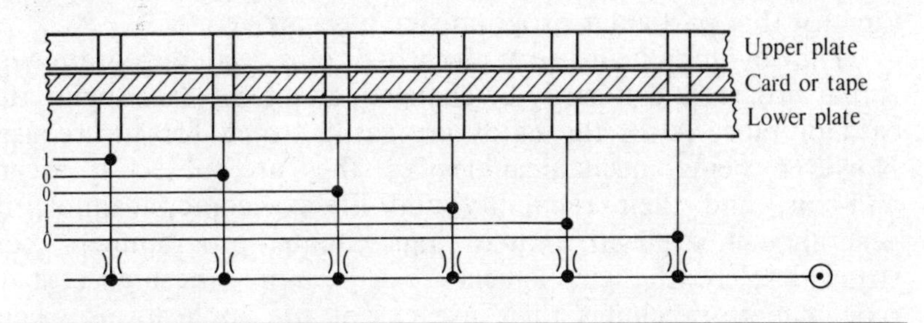

Fig. 8.1 Principle of operation of pneumatic card or tape reader

computer cards. Its output pressure is very low, and the device is intended to be used with turbulence amplifiers.

Pneumatic tape readers operate, in principle, exactly as pneumatic card readers. With tape readers, however, the number of program steps is unlimited, and automatic repetition of the program cycle can be obtained by joining the two ends of the tape. In connection with tape construction and tape advance, two types of tape readers exist:

a. Tape readers utilizing standard punched tape (8 holes per row), with the tape being advanced by a pulse appearing at the 'Advance Tape' terminal.
b. Tape readers using special tapes and a special advance circuit.

Tape readers of the first type require a binary comparator circuit to operate the tape advance. This circuit compares the output signals from the tape reader with the momentary feedback signals arriving from various sensors in the controlled system, and produces an 'Advance Tape' pulse only if there is proper agreement. This binary comparator circuit requires a fairly large number of logic gates, and one such circuit is required for each cylinder or other actuated device. Tape readers operating on this principle are those described in [23] and [24]. The former has 1.4 bar (20 psi) outputs, while the latter operates with lower pressures of 0.1 bar (1.5 psi).

Tape readers of the second type use non-standard wider tapes. Thus, a larger number of output signals become available, without the need for special decoding comparator circuits. The tape reader described in [25] has 20 high-pressure output signals supplying 4–8 bar (60–120 psi). A comparator circuit becomes unnecessary, since up to 20 feedback signals can be accommodated in the tape itself (at 0.3 bar = 5 psi pressure levels), and tape advance only takes place if all the feedback signals entered in the tape for that particular program step have arrived.

The obvious advantage of all card- or tape-reader programmers is that programs can easily be changed, simply by exchanging the card or tape. Also, the cards are easily stored for future use. However, being mechanical devices, they are subject to wear-and-tear, and their reliability and life-expectancy cannot, in general, compare with that of logic-counter programmers constructed solely of logic elements. Furthermore, the high cost of tape readers precludes their use except for applications where frequent program changes are required, and where the number of program steps and of actuated external devices is large enough to fully exploit the capacity of the reader.

2. Stepping drums

Here, a rotating drum is used instead of cards or tapes. The drum is programmed by inserting plastic plugs into holes arranged on the drum in rows and columns. Each row corresponds to a program step, and each column to an output signal connected to some external device. Drum programmers with up to 100 rows and 100 columns are available. As the drum rotates, the plastic plugs inserted in the 'active' row actuate microswitches which then provide the output signals. The location of the plugs can easily be changed by hand.

As in the case of card or tape readers, the drum can be rotated either synchronously, or step-by-step on an event-controlled basis. If the number of program steps is less than the available number of rows on the drum, the programmer must be programmed to 'go home' automatically to Step 1 the moment the last step of the program cycle has been completed.

Cams attached to a rotating shaft are also used instead of drums with plastic plugs. Such cams can supply sufficient force to actuate not only microswitches but also pneumatic or hydraulic valves. On the other hand, making program changes becomes more difficult, and may even involve the need for making a differently-shaped cam. Cam programmers are therefore mainly suitable for more-or-less fixed program cycles.

With stepping-drum programmers, programs can only be stored by removing and storing the drum. The drums, however, are relatively expensive, and also take up storage space. Thus, if a considerable number of programs need to be stored, card- or tape-reader programmers would be preferable. The drawbacks of high cost and of mechanical wear apply here just as much as with card or tape readers, so that the comments made there are applicable here also.

A pneumatic device closely related to the stepping-drum programmer is the 'Bi-Selector', see [26]. Instead of a drum, this device uses a rotor or rotating commutator which is indexed by a stepping cylinder driving a ratchet wheel. The device has 20 input and 20 output connections. The input signals are low-pressure signals (0.8–1 bar, or 12–15 psi), signifying completion of the previous program step. The arrival of the input signal causes the commutator to index to the next position, thereby connecting high-pressure air (3–5 bar, or 45–80 psi) to the next output connection. The device thus provides the mechanical equivalent of a 20-step logic-counter programmer using the $1/n$-code. Unlike the stepping-drum programmer, which is able to produce many output signals simultaneously at each step, the Bi-Selector produces only a single output signal at a time. If more than one device needs to be operated during a given step, and if one of these must be operated again later on during the cycle, then the outputs must be isolated from each other by means of OR gates. (The manufacturer markets a 5-input OR-gate for this purpose.)

If most of the 20 available steps are actually needed for the program at hand, use of the Bi-Selector will be economical. However, if 21 steps should be required, a second Bi-Selector would have to be purchased and coupled to the first one. Furthermore, the Bi-Selector (just as stepping-drum programmers) has a relatively slow switching speed of about 0.1 second (as compared to about 0.001 second for most fluidic elements, and about 0.01 second for moving-part pneumatic logic elements), and this may be a drawback in applications where high production rates are required. An advantage of the Bi-Selector is that simple inexpensive back-pressure sensors can be used instead of limit-valves.

3. *Matrix boards*

A matrix board has a number of rows, each connected to a given input; and a number of columns, each of which is connected to a given output. By inserting plugs into the appropriate intersections, any input can be connected to any one or to several of the outputs. As indicated in Fig. 8.2(a), input row A is, in this example, connected to columns 1 and 4, and input row C to columns 2 and 4. This circuit will not operate properly, since the input signal A will find its way improperly ('sneak path') into

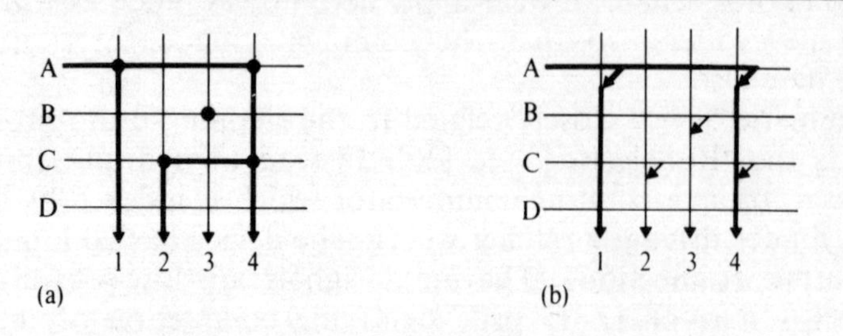

Fig. 8.2 Matrix board (a) with, and (b) without sneak paths

column 2, as indicated by the heavy line. To prevent such sneak paths, plugs with built-in diodes must be used, as shown in Fig. 8.2(b).

While the great majority of matrix board in use are electric, a pneumatic matrix board is also manufactured, see [27], to be used with 2–7 bar (30–150 psi) air pressures.

It must be stressed that the matrix board is not, in itself, a sequential programmer. Its sole purpose is to connect various inputs to various outputs, and to facilitate easy changes of these connections. The matrix board thus functions as an auxiliary device used in conjunction with any type of programmer, whether logic-counter programmer, tape or card reader, or stepping drum, making it more convenient to quickly carry out program changes.

B. Software programmers

The devices to be discussed here use electronic memories to store the required program steps, rather than individual switching elements or the mechanical devices described in the first part of this chapter. A central processing unit, similar to those used in computers, is utilized to read and carry out successive program steps. The memory devices used are byte-organized; i.e., several binary bits in the memory device belong together and form a unit of information called 'byte' or 'word'. One or several such words make up the information required for each program step. While large or medium-sized computers can also be used as sequential programmers, it is less costly to use small-scale computing devices especially designed for this purpose. The so-called programmable logic controllers (P.L.C.) to be discussed first were developed already long before the present micro-processor boom began to take place.

4. Programmable logic controllers (P.L.C.)

The characteristic difference between P.L.C.s and the micro-processors to be discussed in the next section is that the former usually use 1-bit processors; i.e., one single bit is processed at a time. Such relatively simple processors are especially suitable for sequential programmers, but not for carrying out mathematical operations. On the other hand, programming becomes much simpler, and the handling of input and output signals is facilitated,

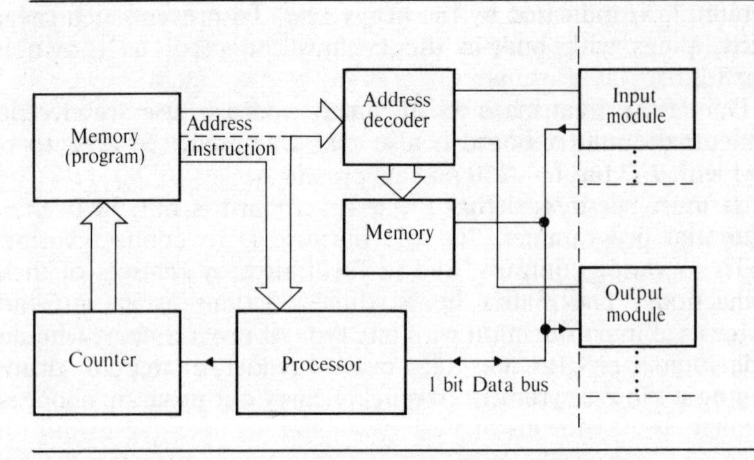

Fig. 8.3 Block diagram of Programmable Logic Controller

with each input or output signal being assigned to a specific address.

The block diagram of a typical P.L.C. is shown in Fig. 8.3 [28]. The basic operation of the P.L.C. is as follows: By means of a counter, the information stored in certain addresses in the memory is recalled, and connected to a number of bus lines. Some of these bus lines lead to the processor, and thus supply the current program instructions to the processor. Other bus lines connect the memory to an address decoder, or even directly to input or output modules. By means of a coded signal, any one of these input or output modules can be selected and connected to the data bus line. Input and output modules for various voltage levels, and for either AC or DC voltages, can be purchased to accommodate different types of external equipment.

As soon as a given processor instruction has been carried out, the counter output is increased by one count, whereby the succeeding program instruction is called in. The time required to process 1K words (= 1024 program instructions) is in the order of 2–5 milliseconds.

The memories used are either of the PROM (programmable read-only memory) or of the Read/Write type. For developing and testing a program, Read/Write memories are more convenient. These frequently use magnetic-core memory units, since magnetic cores do not lose the information stored in them because of power interruptions. Lately, the use of semiconductor RAM (random-access memory) Read/Write memories has become more popular, since these cost much less than magnetic core memories. These are operated in conjunction with an emergency battery power supply, which takes over in case of a power interruption, and thus prevents loss of the information stored in the memory.

Once the program has been tested, it can remain in the Read/Write memory and used there, or transferred to a PROM memory, provided no future program changes are expected. PROM memories are less sensitive to accidental erasure or to electrical disturbances.

The programming of P.L.C.s is carried out by means of a special programming unit. Such units are usually expensive, but are required only during the actual programming. Thus, a plant employing several P.L.C.s need only purchase a single programming unit. The program is entered not in the form of any special language, but by means of Boolean equations or in the form of a ladder diagram. The latter method is a carry-over from relay circuits. Operating personnel in the habit of working with relay circuits feel at home with ladder diagrams, and prefer to use such diagrams to express the program requirements, even where relays are not used anymore.

Figure 8.4 illustrates a typical programming procedure using Boolean equations. In the example, the operation $x_7 = (x_2 + x_3 + x_8)x_1'$ is carried out, and the resulting value of x_7 stored. Note that each one of the inputs is called in from its memory address sequentially, unlike regular logic circuits in which all inputs arrive simultaneously.

The program size that can be accommodated depends on the capacity of the P.L.C. and this, in turn, depends to a great extent on its price. Different manufacturers offer P.L.C.s with different capacities, and the capacity is usually expressed in terms of

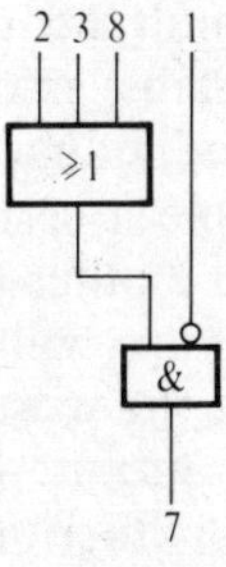

Program		Comment
Instruction	Address	
O	2	O = OR
O	3	AC = AND compliment
O	8	
AC	1	S = Store
S	7	

Fig. 8.4 P.L.C. programming using logic symbols

memory words available. However, the number of words required to program a particular program depends, to a great extent, on the programming method used. This should be carefully checked before selecting a given P.L.C. Many P.L.C.s also provide additional features, such as counters and timer units.

P.L.C.s were first made available about 10 years ago, and are now being manufactured by many different companies [29], [30], [31]. Their prices were lowered considerably at first, but have by now more or less stabilized. The least expensive P.L.C.s, having about 10 input and 10 output modules, cost in the order of $2000 (£1000), including the programming unit. More elaborate P.L.C.s cost four of five times that sum. In applications where pneumatic devices (cylinders, valves, etc.) must be actuated, additional money must be spent on solenoid valves for transforming the electronic outputs of the P.L.C. into suitable pressure signals. It is obvious that the P.L.C. cannot compete with logic-counter programmers on an economic basis, except for very complicated programs, where the full capacity of the P.L.C. is being utilized.

5. *Microprocessors*

Microprocessors cost only a fraction of P.L.C.s, and their price is continuing to drop. However, they are relatively difficult to program, and thus do not provide anything near the operating convenience of programmable logic controllers.

Microprocessors operate with a specific word length, usually 4, 8 or 16 bits [32] and [33]. This is the number of binary digits which are entered, processed and leave the processor simultaneously. Here, already, the difference between microprocessors and P.L.C.s makes itself felt. Microprocessors are generally able to carry out addition and subtraction, and frequently also multiplication and division. Several inputs are sampled simultaneously (e.g., 8 inputs for an 8-bit processor). In many available microprocessors, it is usually not possible to sample an individual bit within a word. For example, if it is desired to sample a certain input, this input will enter the processor together with 7 other bits (again, for an 8-bit processor). Additional operations will be required to obtain the value of the required bit within the word.

Similar problems appear in connection with the output. For instance, if we wish to call for a specific output bit, which may be either 0 or 1, we must also determine and call for the values of all the other bits belonging to the same word. It thus becomes clear that 1-bit processors are more suitable as sequential programmers than word or byte-processors.

There are systems utilizing more than one processor: a bit processor is used as a sequencer, and works in conjunction with a byte-processor used for mathematical operations, evaluating instructions, etc. Such systems usually make use of subprograms for realizing logic functions (AND, OR, NOT, etc.), so that the programming of the required sequence becomes relatively simple. For the user, it may not even be apparent whether he is working with a bit- or byte-processor.

APPENDIX

CLASSIFICATION OF COMMERCIALLY AVAILABLE ELEMENTS

In order to make this book of more practical use to the designer, a list of some manufacturers supplying the various types of logic elements is given below. This list makes no pretence of being complete, and nothing concerning the quality of the respective elements is meant to be implied because of the inclusion or omission of any name.

Category A: Miniature pneumatic 3/2 or 5/2 valves with double-pilot actuation

Aro (USA) – 5/2 valves
Athos-Leibfried (West Germany) – 3/2 and 5/2 valves
Atlas Copco (Sweden) – 5/2 valves
Bosch (West Germany) – 5/2 valves
Clippard (USA) – 3/2 and 5/2 valves
Crouzet (France) – 5/2 valves
Dynamco (USA) – 5/2 valves
Festo M5 System (West Germany) – 5/2 valves
Gachot (France) – 3/2 valves
Herion (West Germany) – 5/2 valves
Joucomatic (France) – 3/2 and 5/2 valves
Kay (England) – 3/2 and 5/2 valves
Kuhnke (West Germany) – 3/2 and 5/2 valves
Lang (England) – 5/2 valves
Martonair (England) – 5/2 valves
Maxam (England) – 5/2 valves
Miller (USA) – 5/2 valves
Numatics (USA) – 5/2 valves
Schrader (England) – 5/2 valves
Telemecanique (France) – 3/2 valves
Wabco Westinghouse (USA) – 5/2 valves

Category A also includes all manufacturers of conventional fluid-power valves, but, because of their great number, no attempt has been made to list these here.

Category B: Systems of moving-part fluidic elements providing basic switching functions but no flip-flip element

Festo System 1000 (West Germany)

Category C: Moving-part fluidic elements providing only a single complex switching function

Manufacturer	Function
Dreloba (East Germany) GEC-Elliott (England)	$F = AC + B'C + A'D + BD$
Samsomatic (West Germany) Genicon (USA)	$F = AB'C + D(A' + B)$
Hoerbiger (Austria)	$F = AB + A'C$

Category D: Moving-part fluidic or pure fluidic elements providing only the NOR function

Double-A (USA) – Moving-part fluidic, fan-in = 3
Sempress (Holland) – Moving-part fluidic, fan-in = 4
Techne (England) – Moving-part fluidic, fan-in = 2
Johnson (USA) – Impact modulator, fan-in = 4
Asco (USA) – Turbulence amplifier, fan-in = 4
Agastat (USA) – Turbulence amplifier, fan-in = 4
Martonair (England) – Turbulence amplifier, fan-in = 2

Category E: Wall-attachment fluidic elements

Corning (USA)
Knorr-Bowles (West Germany)

Manufacturers supplying standard flip-flop elements with AND gates in the SET and/or RESET inputs:

Bosch (West Germany) – (AND gate only in SET input)
British Fluidics and Controls (England)
Contraves (Switzerland)
Norgren (USA)

REFERENCES

1. Huffman, D. A., 'Synthesis of sequential switching circuits', *J. Franklin Institute*, Vol. 257, No. 3 & 4, 1954.
2. Cole, J. H. and Fitch, E. C., 'Synthesis of fluid logic control circuits', *Proc. Joint Automatic Control Conference*, Aug. 1969, Boulder, Colorado, pp. 425-32.
3. Cole, J. H. and Fitch, E. C., 'Synthesis of fluid logic circuits with combined feedback input signals', *Fluidics Quarterly*, Vol. 2, No. 5, 1970, 14-21.
4. Cole, J. H. and Fitch, E. C., 'Synthesis of fluid logic networks with optional input signals', ASME Paper 70-Flcs-19.
5. Knoerr, B., 'Universal application possibilities of moving part elements', *Proc. Seventh Cranfield Fluidics Conference*, Nov. 1975, Stuttgart, Paper H2.
6. Goehring, D., 'Fluidically controlled filling unit', *Proc. Seventh Cranfield Fluidics Conference*, Nov. 1975, Stuttgart, Paper H4.
7. Schmidtke, W., 'Application of turbulence amplifiers in industrial circuitry', *Proc. Seventh Cranfield Fluidics Conference*, Nov. 1975, Stuttgart, Paper H5.
8. Pessen, D. W. and Golan, G., 'Do-it-yourself programmable controllers', *Hydraulics & Pneumatics*, Oct. 1975, pp. 182-7.
9. Vingron, P., 'An assembly technique for sequential circuits', *Proc. Fifth Cranfield Fluidics Conference*, June 1972, Uppsala, Sweden, Paper F6.
10. Cheng, R. M. H., 'Application of fluidic shift-register modules for sequential control of pneumatic sequential circuits', *Proc. Fifth Cranfield Fluidics Conference*, June 1972, Uppsala, Sweden, Paper F2.
11. Technical Handbook – Telemecanique (TE controls).
12. Fasol, Hübl and Vingron, 'Statische pneumatische Logiksysteme', Verlag Berliner Union-Kohlhammer, 1972, pp. 171-6.
13. Hübl, Fricke and Streppel, 'Entwurf sequentieller Steuerungen in der Pneumatik', 2. *Aachener Fluidtechnisches Kolloquium*, 1976, Vol. 2, 33-38.
14. Backé, W. and Goedecke, W. D., 'Pneumatische Schaltungstechnik', R. W. Technischen Hochschule Aachen, 1975, pp. 75-88.

15. Festo System 1000 Catalog, April 1975, pp. 102-10.
16. GEC-Elliott Dreloba Standard Einheiten, Sept. 1975, pp. 8-9.
17. Hübl, W., 'Ein Beitrag zur Synthese asynchroner Grund- und Standardschaltwerke', *Schriftenreihe des Lehrstuhls für Mess- und Regelungstechnik*, Heft 5, Ruhr-Universität Bochum, 1975.
18. Bulletin 1000 'Program-Air', Dynamco Inc., USA, May 1976.
19. Industrial Timer Corporation, USA.
20. Automatic Timing & Controls, Inc., USA.
21. Techne Ltd., England.
22. ASCO Fluidics, USA.
23. GEC-Elliott Dreloba.
24. Festo, West Germany.
25. Atlas-Copco, Sweden.
26. Martonair, England.
27. Gachot, France.
28. Motorola, Industrial Control Unit MC 14500B.
29. issc, industrial solid state controls, USA.
30. Allen Bradley, USA.
31. Siemens, West Germany.
32. Intel, 8080 and 8085, USA.
33. Texas Instruments, TMS 9900, USA.

INDEX